HARMONIZING EUROPE

SUNY series in Global Politics
James N. Rosenau, editor

HARMONIZING EUROPE

Nation-States within the Common Market

Francesco G. Duina

STATE UNIVERSITY OF NEW YORK PRESS

Published by
State University of New York Press, Albany

© 1999 State University of New York

All rights reserved

Printed in the United States of America

For information, address State University of New York Press,
State University Plaza, Albany, N.Y., 12246

Production by Cathleen Collins
Marketing by Dana Yanulavich

Library of Congress Cataloging in Publication Data

Duina, Francesco G., 1969–
    Harmonizing Europe : nation-states within the common market /
Francesco G. Duina.
        p.   cm. — (SUNY series in global politics)
    Includes bibliographical references and index.
    ISBN 0-7914-4177-6 (alk. paper)
    ISBN 0-7914-4178-4 (pbk. : alk. paper)
    1. Law—European Union countries—International unification.
2. Equal pay for equal work—Law and legislation—European Union
countries.   3. Air—Pollution—Law and legislation—European Union
countries.   4. Equal pay for equal work—European Union countries.
5. Air quality management—European Union countries.   I. Title.
II. Series
    KJE971.5.D85   1999
    337.1′4—dc21   98-30737
                                                                  CIP

10 9 8 7 6 5 4 3 2 1

To my parents, Liliana and Gianfranco

Contents

Contents

Tables

Foreword

It is always disturbing to see fashion, whose life-blood is that of movement between extremes, dominate academic life. Such domination has certainly characterized studies, particularly those undertaken in North America, of the various stages of European integration. It takes little effort to recall the move from an early functionalist account of purportedly inevitable evolution toward political unity to gloomy accounts of "eurosclerosis." The unification of Germany and the Maastricht Treaty have again brought optimism to the fore. This is a matter of some consequence for social understanding in general: the presumption that Europe will become a genuinely transnational entity is, for example, much cited by the "globalizers" (who are indeed everywhere), that is, by those who seek to convince us that the era of the nation-state is coming to an end. However, one notes already, in the wake of skepticism about the launching of the euro, the emergence of a renewed strain of pessimism—itself less the replacement of hope by analysis than an example of the pleasure so often taken in prophesying doom.

No point of investigation could be better judged to advance genuine understanding than that of an examination of the extent to which the different European states adopt the directives of the European Union. Francesco Duina's masterly study of two such directives accordingly does more to advance knowledge of what is happening in Europe than does a very great deal of the extant writing in the field. The famous remark of Aby Warburg—that "the good Lord lives in details"—most certainly applies here. Duina's microscope reveals a huge amount.

To begin with, the differential rates of implementation of these directives demonstrate beyond any question that the nation-state in Europe is most certainly not dead. Duina has much to say at this point. He might well accept the sensible and brilliantly argued view of Alan Milward to the effect that the European Union was created by statist design, above all that of France seeking to control Germany, and that it has done much to preserve and even solidify

Europe's nation-states.¹ But agreements reached by leaders of states in the international frame of Europe nonetheless involve pressure to rationalize the institutions of national societies. So the situation is complex and messy, witnessing pressures to harmonize in order that states can survive and prosper. No manichean either/or division between a (mythical) all-powerful nation-state and transnationalism can do justice to this situation: what we witness instead is the emergence of modest nation-states, giving up some powers because doing less is the best route to achieving more.

That this book is quite as much a major contribution to the comparative study of state-society relations can be made clear by highlighting the analytic tools that are used by the author. Great sophistication is shown in the way in which the mere acceptance of a directive is distinguished from its implementation. Ruling bodies make all sorts of claims to power, but this author is well aware that what matters is their ability to find agents to translate ideas into practice. Then we are offered marvelously succinct analyses of the way in which the legacies of national institutions constrain and channel later policy-making. In this area, it is important to note the author makes a decisive argument, distinguishing himself from the "historical institutionalists" with whom he otherwise has so much in common. He finds that states have far less autonomy than many scholars have recently asserted. Here too we are offered a modulated and convincing portrait of the genuine complexities of state-society relations in developed capitalist societies. One of the peculiarities of the European Union is that convincing predictions about the future are hard to come by. Progress has been fitful, very much a matter of apparent blockages leading to surprising breakthroughs. This welcome contribution to the literature offers us the tools to understand both sides of this equation. It is the best guide to the future of the European Union known to me.

Professor John A. Hall
McGill University
Montreal, Canada

Acknowledgments

I would like to thank my dissertation committee members at Harvard University's Department of Sociology, Theda Skocpol, John Campbell, and Yasemin Soysal, for their guidance, support, time, and commitment during the early stages of this book. I would also like to thank Professor John Hall for his generous help, attention, and time, and Special Advisor to the European Commission Renée Haferkamp for her advice. I am grateful to my colleague John Glenn for reading several drafts of this manuscript and continuously challenging me with stimulating questions; Emil Dabora for his support of this project and advice over the years; Adnan Afridi, Andrew Blom, and Russell Epstein for spending time on the consistency of my arguments. I am also indebted to my wife Angela Atkinson-Duina for her patience and support, and to my brothers Andrea, Alessandro, and Pierluigi, and my friends Patrick Lohier, David Schab, Brad Thompson, Lisa Ceglia, Dan Pikelny, and Raiko Mancini for their editorial work and support. Finally, I would like to thank Harvard's Center for International Affairs, Harvard's Center for European Studies, Harvard's Graduate Society, the Leopold Schepp Foundation, and the Westengard Foundation for their financial support and my colleagues at the Monitor Company for kindly allowing me to take time off to complete this book.

Abbreviations

CBI	Confederation of British Industry
CEGB	Central Electricity Generating Board
CFDT	Confédération Française Démocratique du Travail
CGIL	Confederazione Generale Italiana del Lavoro
CGL	Confederazione Generale del Lavoro
CGT	Confédération Générale du Travail
CIMA	Comisión Interministerial del Medio Ambiente
CISL	Confederazione Italiana dei Sindacati dei Lavoratori
CSAS	Coal and Smoke Abatement Society
CTF	Comité du Travail Féminin
DC	Democrazia Cristiana
DPCM	Decreto del Presidente del Consiglio dei Ministri
DPR	Decreto del Presidente della Repubblica
EC	European Community
ECR	European Court Report
EEC	European Economic Community
EOC	Equal Opportunities Commission
EU	European Union
EPA	Equal Pay Act
EPD	Equal Pay Directive
ILO	International Labour Organisation
MDF	Ministère des Droits de la Femme
PCI	Partito Comunista Italiano
SSD	Sulphur Dioxide and Suspended Particulates Directive
TUC	Trade Union Congress
UIL	Unione Italiana del Lavoro

1

Introduction

Harmonizing Europe with Directives

Most of the world is currently witnessing the spread of capitalist markets. In some areas, like Southeast Asia and Russia, this is primarily a phenomenon at the level of nation-states, as nations struggle to build the necessary institutions. In other areas, like North America and Europe, it is an effort at the transnational level, as politicians and policy-makers gather to agree on rules and regulations, such as the North American Free Trade Agreement, that can foster international trade among participating states.

My objective in this book is to demonstrate, through the example of the European Union (EU), that the rules and regulations used to build transnational markets consistently demand dramatic *institutional*, and hence *cultural*, transformations in member states, and to argue that those demands explain why transnational markets stall when they do. Capitalism, as Max Weber once wrote, requires "the absence of irrational limitations on trading in the market" (Weber 1992, 276). No one doubts that transnational markets must be free of irrational limitations, such as national differences in environmental constraints on industry; few, however, truly understand what the removal of those limitations entails.

Transnational markets challenge the institutions of member states in very specific and tangible ways. By introducing rules and regulations that member states are expected to ratify or adopt, they pose a direct challenge to existing *national legal systems*. National legal systems are, however, much more than a set of abstract rules: they are the answers that members of society have surmised, at times of specific political and economic conditions, to deal with the difficulties of social life. They hence reflect deeply rooted, collective values and beliefs, interpretations of problems, issues, and life. Great Britain's understanding of "air pollution" and how to control it legally in the 1980s, for instance, was the product of the timing of the industrial revolution, the

1

perceived sacredness of domestic fires, the available technology, and the political compromises of the 1800s between industrialists and the state. An European Community (EC) directive with unprecedented, apparently merely technical rules meant for Great Britain a profound transformation of long-standing and widely accepted principles and ideas.

Legal systems regulate societies through administrative structures that apply their principles. Accordingly, when transnational markets demand changes in laws, they also demand changes in *administrative structures*. These structures, however, are as rooted in history as the laws they are designed to apply. In the 1980s, the decentralized, informal administrative apparatus for controlling air pollution in Great Britain descended from the historical legacies of the nineteenth and eighteenth centuries. An EC directive demanding the centralized, uniform application of new principles of air pollution control asked in effect for an administrative revolution.

The challenges of transnational markets reach beyond the legal and administrative legacies of nations. They also threaten the *distribution of resources and goods among groups* in society. Within the boundaries of nation-states and their legal systems, interest groups have learned to adapt, accepting existing limitations while investing heavily in safe practices. Industrialists in Great Britain, for instance, learned through time that paying women far less than men for different jobs, even when these could be deemed of equal value, was a perfectly acceptable way of keeping labor costs down and boosting their competitiveness within their country and abroad. The British state, after all, had resisted drafting equal pay legislations since the 1800s. The arrival of an EC directive on equal pay for men and women in 1975 threatened to weaken dramatically two interest groups—British capital and working men—while enriching a third, working women, that for very long had been terribly weak.

Nation-states have answered the challenges of transnational markets with concrete measures. While their representatives have committed themselves to the principles of the new agreements, domestic legislators and policy-makers have found it difficult to impose on society the spirit and application of principles that overrun long-standing norms and traditionally powerful groups that have grown out of particular legal environments. Systematically and consciously, they have failed to ratify or properly import basic legal concepts, belatedly miscommunicating goals and objectives or outright ignoring the new rules. They have operated with the perceived well-being of the nation in mind, and inevitably under pressure from strong interest groups alarmed at the possibility of losing power.

It is impossible for transnational markets to avoid challenging the legal foundations and the powerful interest groups of member states. They require, in order to function, national legal systems and interest groups subject to identical constraints. Yet, as this book demonstrates with the European Union, it would be wrong to claim that these challenges concern every member state at any given time. In some instances, the demands of markets match the legal and administrative contexts of certain nation-states, or they reinforce the current distribution of resources among groups in society. In such nations, the new principles are respected and upheld.

Similarly, it would be wrong to claim that the demands of transnational markets are doomed to encounter complete rejection anywhere they demand deep transformations. More commonly, controversial principles find their way even in the more recalcitrant of the member states but only partially and belatedly. National idiosyncrasies and international mandates then coexist in contradiction with each other, both under pressure to change. It follows that I do not intend to argue for the ubiquitous rejection of the transformative demands of transnational markets, but rather for an appreciation of the fact that transnational markets do demand transformation and that, as a result, they are bound to encounter resistance and therefore experience differences in implementation across member states.

I will prove my argument by turning to the European Union,[1] a fifty-year-old struggling but major experiment to merge the economies of several powerful capitalist nations (full compliance with EU law is set to around 50%), and I will concentrate on the demands and fate of directives, the primary tool for creating a single market, in member states. I will show how directives, despite their apparent technical content, have challenged national institutions and hence the fundamental values of some member states, and thus why their implementation has varied across the Union.

My hope is that, by reaching deep into the demands and fate of directives, I offer an unprecedented, full picture of what transnational capitalism really asks, albeit only for the European Union. There exists a growing literature on transnational markets, their demands, and the resistance of entrenched national institutions and interest groups to changes in laws and resources. Vogel (1995), for instance, discusses in detail the environmental demands that the EU, GATT, and NAFTA have posed on member states and the resulting reaction from domestic interest groups and legislators. Keohane and Milner (1996), Katzenstein (1985), and Golden (1986) have, in a similar spirit, analyzed the resistance of and the relationships between domestic institutions and transnational, as well as international, markets. None of these works, in my view, offers a *systematic* theory of what, exactly, transnational markets ask of member states, and *illustrates* and tests its assertions with a systematic, in-depth examination of the fate of those demands in domestic settings. My hope is thus to complement the existing literature by offering a clear analytical theory of the demands of transnational markets accompanied by an exhaustive empirical illustration of that theory.

By applying my theory to the EU, I also outline what I think is one of the first theories of implementation in the EU to date, despite the poor implementation record of member states and the significant number of scholars working on the EU.[2] Indeed, most literature on the EC has covered various aspects of policy-*making*, rather than implementation.[3] When focused on implementation, past works have documented implementation patterns, rather than offer an explanation for them. Thus, Siedentopf and Ziller (1988) openly state that "the objective of [their] comparative review is an overview presenting facts and examples given by national reports" on implementation and not an explanation of the facts themselves. Their work is otherwise very comprehensive: it covers seventeen directives from different policy areas in ten nations. Bennett (1991) reports on the implementation of air pollution control directives in the twelve member states as of 1991. A number of facts are used to describe the transposition, application, and impact of directives, but no theory is offered to explain outcomes. Landau (1985) and Warner (1984) provide a detailed account of the impact of Article 119 of the Treaty of Rome on equality of pay, social security, and equal treatment, but no explanation of what limited or drove that impact. Vogel-Polsky (1985) studies the types of institutional and noninstitutional structures that have grown in member states to implement gender legislations. This work is echoed by those of Brewster and Teague (1989) and Mény et al. (1996), when they consider the impact of the EU policy on national institutions and policies. None of these works, however, outlines a theory of implementation.

The Case of the European Union

The European Union dates back to the years following World War II and unites, through several treaties and agreements, fifteen economies, including three of the five most powerful in the world (Germany, France, and Italy). It began in 1951 in Paris with the creation of the European Coal and Steel Community (ECSC) among France, Germany, Belgium, Italy, the Netherlands, and Luxembourg. In 1957, the same six signed the two Treaties of Rome: the European Economic Community (EEC) and the European Atomic Energy Community (Euratom), formally giving birth to what was known, until 1993, as the European Community (now European Union) (Nugent 1991, 42).[4] At different stages, nine more members joined, as additional treaties, acts, and charters, including the Single European Act of 1986 and the Maastricht Treaty of 1992, sought to add or specify in greater detail the common objectives of European member states and to accelerate the elimination of those 'irrational' barriers, to use Weber's terminology, that precluded the existence of an efficient single market.

Powerful laws, known as directives, have articulated the principles of each major treaty into binding words.[5] To comply with directives, the parliaments of member states have had to draft laws whose principles superceded those of any existing legislation (a process I refer to as the *transposition* of directives),[6] while national administrations have had to mobilize to ensure the application of the new principles, regardless of whether the appropriate infrastructure existed or whether existing policies or structures conflicted with the new mandates (a process I call the administrative *application* of directives).[7] Theoretically, the exact wording of the new domestic laws and the administrative means for application were left to the discretion of members states, as Article 189 (3) of the EEC Treaty specified:[8]

A directive shall be binding, as to the result to be achieved, upon each Member State to which it is addressed, but shall leave to the national authorities the choice of form and method.

Practically, nothing less than a faithful, almost word by word, transposition of the directive into national laws and the choice of the "most appropriate forms and methods" of administrative application were expected (Capotorti 1988, 159), with the European Commission and the European Court of Justice empowered with the right to judge and punish transgressors.[9]

Directives have dealt mostly with the economic sphere, such as the free movement of persons, capital, and services, the erection of common external tariffs, and the introduction of commercial rules, but the transnational market has required that other policy areas find a place on the Community's agenda. The early EEC Treaty provided the basis for a Community Social Policy in Articles 117–22, and a Social Fund in Articles 123–28. Later, more elaborate directives over gender equality, immigration, education, workers' rights, and labor markets were produced, culminating with declarations such as *The Community Charter of Fundamental Social Rights for Workers* written by the Commission in 1989.[10] Most recently, energy, research and technological development, environmental, sectoral (e.g., fishing, steel), and foreign relations have all been regulated, leaving "few policy areas with which the [Union] does not have at least some sort of involvement" (Nugent 1991, 272).

Not all directives have targeted major sectors of the economy, society, or the polity. Some have merely set rules in areas of little relevance or have affected a handful of persons or groups. Others were formulated so loosely that their transposition into national law challenged little.[11] By and large, however, directives have been the major tool used to build the European transnational market. Hence, we must turn to them when wondering what transformations, really, the European transnational market has asked of member states.

In the chapters that follow, I intend to demonstrate that directives have challenged the legal-administrative legacies of member states and the interest groups that grew in those legal-administrative contexts. Normally, any one directive has corresponded to the legal-administrative landscape of some member states, and, accordingly, has left interest groups on safe grounds. Yet, inevitably, the same directive has demanded deep transformations in some other member states. It was its nature to do so, for the creation of the single market entailed precisely the harmonization of differences *among* the participants. I intend to show that legacies and groups have emerged from decades or centuries of social, political, and economic struggles, embodying collective consciousnesses and interests. Directives, that is, have challenged deeply rooted social entities. This, I will argue, explains why any one member state has only selectively implemented EC and EU directives: only those directives that have affirmed existing ideas and interests, or that left them alone, have found their way in the legal systems of a member state.

When the challenges were great, I will show, parliaments stepped up as the guardians of the status quo, as the shield protecting their countries against radical demands descending from the EC and the EU. Legislators proved unwilling to build the consensus necessary for drafting transformative laws. Similarly, they barred the creation of new administrative structures, changes in the informal and formal operating procedures within and outside of the administration, and the allocation of resources to new, unprecedented tasks. Parliaments accepted only those directives that were in consonance with past policies.[12]

At the same time, legislators, unable to upset existing economic and social systems, opposed laws that they knew would undermine the current organization of interest groups. Interest groups, in turn, merely reinforced the decisions of legislators: strong groups supported the state's protection of the status quo, while weak groups mobilized, generally unsuccessfully, to pressure the state to adopt a radical directive or to oppose the implementation of a directive that merely reaffirmed the status quo.

The outcome of this protectionism has been the *variation* in implementation of any given directive that one can easily observe across the European Union. In most cases, either the *timing* of transposition or application, or the *extent* of transposition or application of a directive has been faulty in one or more member states. Member states have been *late* in transposing or applying the directive's principles,[13] or have *failed* to transpose all of the principles found in the directive (with modifications in numbers, deadlines, concepts, affected parties, regions, and sectors of industries) or to apply those principles (through the imposition of weak penalties, sanctions, etc.).[14]

Table 1.1 summarizes this theoretical viewpoint, without addressing explicitly the historical origins of interest group strength and policy legacies

Table 1.1. The Institutional Explanation

Demand on Domestic Institutions	*Mechanisms*	*Impact*
• Major shift in policy legacy	• **State opposes directive**	• National version reflects existing laws and administrative traditions; elimination of principles that depart from status quo
• Major reorganization of interest groups	• **State opposes directive** • Strong interest groups oppose directive • Weak interest groups cannot ensure implementation of directive	• National version void of any principle that undermines strong interest groups; santions weak from the start
• Consistent with policy legacy	• **State supports directive**	• National version fully embodies the principles of the directive; drafting is punctual and administrative measures are serious
• Consistent with organization of interest groups	• **State supports directive** • Strong interest groups support the directive • Weak interest groups cannot block implementation of directive	• National version fully embodies the principles of the directive; drafting is punctual and administrative measures are serious

that explain why domestic institutions endure in spite of transnational pressures. Naturally, not all directives have challenged both the state policy and interest groups of any one member state. Some directives have been in consonance with parts (typically the legal tradition) of existing policy legacies and, at the same time, have imposed major changes on interest groups. Textual transposition of such directives into national law, without the actual application, has then been typical. Other directives have challenged parts of existing policy legacies but have been perfectly in line with existing interest group conditions. Those directives have then been transposed without application or simply applied without being transposed.

My argument may seem to leave little hope for the completion of a single European market. If I am correct, the European transnational market will not take place, since directives cannot eliminate institutional differences between countries. But putting the problem in such strong terms is by no means intended to dismiss the European effort as futile. The European Union has not completely failed to transform member states' legal-administrative systems. Even in the case of the worst offenders, such as Italy and Belgium, convergence toward the single market is an accepted, obvious fact of life. A fair percentage of the most challenging directives is implemented in all member states, however belatedly and partially.

It is, then, rather a question of *degrees* of convergence, of resistance, of the co-existence of enduring national idiosyncrasies and transnational elements; it is a question of appreciating and understanding that national institutions and cultures are being eradicated and replaced with *standardized answers* to the problems of social life; and, finally, it is a question of comprehending the reasons for countries' failures to implement fully and on time.

Moreover, by painting a clear and realistic picture of the functional state of the Union, I intend to direct policy-makers toward more constructive strategies for unification. By identifying dynamics hitherto unknown concerning the implementation of directives, the book offers initial guidance on how to improve implementation in the future. The conclusion suggests constructive ideas for future planning, such as assigning a stronger role to the European Court of Justice, improving the design of directives, and carrying out consultations with national interest groups and legislators.

I chose to illustrate my thesis by way of in-depth examination of the fate of two directives in selected member states. Numerous directives could be used for this purpose, but I have selected directive 75/117EEC on Equal Pay and directive 80/779EEC on Sulphur Dioxide and Suspended Particulates (Smoke) in the Air. I selected these directives for the following reasons: their period for implementation has fully elapsed and some years have passed since the deadlines expired, allowing for the possibility of observing not only member states' activities during the allowed time, but also their reactions when the European Commission pressured unobserving members to adopt those laws; the directives, secondly, showed great degrees of variation in implementation across member states, making it imperative that any explanation I offer be able to account for the failures but also successes of the EU; and they involved, by virtue of being controversial, burdensome, and potentially far-reaching, a number of actors and structures, making it less likely that any single explanation can truly make sense of the turn of events. The two directives, moreover, regulated what I felt were fascinating areas of social life, ones that touched in one way or another the lives of most citizens of the then Community.

The Equal Pay Directive (EPD) introduced the principle of equal pay for work of equal value and abolished discriminatory clauses in collective agreements. Its aim was to maximize the use of human capital and to promote gender equality in the labor market. It mobilized and affected trade unions, women's groups, governmental departments and ministries, and employers' organizations. Wherever implemented, it matched women's skills and capacities to commensurate rewards, thus promoting the ideal use of female labor and bringing women one step closer to real equality at the workplace and in society.

The Sulphur Dioxide and Suspended Particulates Directive (SSD) introduced standards for air quality in order to hold all industrial producers to identical standards and as a major attempt to fight air pollution throughout the member states. Introduced in 1980, it affected industries, power plants, and all producers of exhaust. When implemented, it cleaned the air from one of its worst polluters in history.

The implementation of the EPD is considered for France, Italy, and Great Britain; that of the SSD for Italy, Great Britain, and Spain. There, where a host of differences beyond institutional arrangements separate these countries, the directives experienced the most variation, making it less likely, again, that a single explanation can account successfully for the outcome of events.

With the theoretical underpinnings of my arguments thus elucidated, let me describe the content of the following chapters. Chapters 2, 3, and 4 analyze and account for the implementation of the EPD in England, Italy, and France respectively. Chapter 5, 6, and 7 analyze and account for the SSD in Italy, Great Britain, and Spain respectively. Chapter 8 compares the main findings for the different case studies, considers possible alternative explanations for the outcome of events, discusses the implications of the findings for the future of the European Union and transnational markets, and suggests venues for future research.

Part I

The Equal Pay Directive

Prologue: The Directive

The Equal Pay Directive (EPD), adopted by the EC Council in 1975,[1] was one of the earliest and most decisive proofs of the thoroughness with which the Community was trying to build a single European market. The Community was now beginning to remove, to put it in Weberian terms, 'irrationalities' in the labor market, issues traditionally ignored in the 1950s and 1960s.[2]

Article 119 of the 1957 Treaty of Rome, with its narrow notion of 'equal pay for equal work', had done little to ensure gender equality. The European labor market, highly segregated along gender lines, made the Article applicable only to a small number of cases.[3] Europe needed a far more powerful law, one that could enable women to claim equality of pay for different work, if it wished to eliminate unyielding wage differentials between men and women.[4]

The first and most important requirement of the directive, therefore, was that men and women receive equal pay for equal work or 'for work of equal value'. Article 1 extended the principle of equal pay outlined in Article 119 of the Treaty of Rome to work to which equal value could be attributed:

> The principle of equal pay for men and women outlined in Article 119 of the Treaty, herein called 'principle of equal pay', means, for the same work or for work to which equal value is attributed, the elimination of all discrimination on grounds of sex with regards to all aspects and conditions of renumeration.

The second requirement followed naturally from the first. For the first requirement to have a real impact on labor markets, each country was expected to nullify "collective agreements, wage scales, wage agreements or individual contracts of employment . . . contrary to the principle of equal pay" (Article 4). This same notion was expressed earlier in Article 1: wages "must

11

be based on the same criteria for both men and women and so be drawn up as to exclude any discrimination on the grounds of sex."

Thirdly, pay had to be defined broadly, as in Article 119, so that employers could not discriminate by ways of allocating forms of compensation other than wages differently between men and women. The directive thus defined pay to be "the basic or minimum wage or salary and any other kind of consideration, whether in cash or in kind, which the worker receives, directly or indirectly, in respect of his employment from his employer."

The timing allowed for the transposition was one year:

> Member states shall put into force the laws . . . necessary in order to comply with this Directive within one year of its notification. (Article 8)

The directive offered only broad guidelines on how member states should ensure proper application. Article 2 simply stated that:

> Member states shall introduce into their national legal systems such measures as are necessary to enable all employees who consider themselves wronged by failure to apply the principle of equal pay to pursue their claims by judicial process after possible recourse to other competent authorities.

Article 6, similarly, stated that:

> Member states shall, in accordance with their national circum-stances and legal systems, take the measures necessary to ensure that the principle of equal pay is applied. They shall see that effective means are available to take care that this principle is observed.

Only Article 7 asked that information concerning the law be brought to the attention of all employees by "all appropriate means, for example at their place of employment."

While relatively free in their choice of applicative means, member states were expected to be capable of applying the directive's principles within one year (Article 8). Table I.1 summarizes how France, Italy, and Great Britain faired on the transposition and applicative fronts. France and Italy transposed the directive rather quickly. Although belatedly, France was alone in setting up the relevant administrative structures that could apply the EPD's principles. Great Britain, on the other hand, resisted implementation: the directive was only partly and belatedly transposed, and never applied. Such dramatic variation in implementation, rather typical of EU directives in general, pre-cluded the smooth construction of a single market in which female labor could be fully recognized and utilized. To put it again in Weberian terms, the EPD's fate in Europe was proof that the removal of irrational barriers, so

Table I.1. Implementation Records of France, Italy, and Great Britain for Directive 75/117

75/117 Required:	Equal Pay for Work of Equal Value	Nullification of Discriminatory Agreements	Broad Definition of Pay
France:	Transposed fully on time, applied almost fully but late	Transposed fully on time, applied almost fully but late	Transposed fully on time, applied almost fully but late
Italy:	Transposed fully almost on time, poorly applied	Transposed fully almost on time, poorly applied	Transposed fully almost on time, poorly applied
Great Britain	Transposed in part and late, never applied	Transposed fully on time, never applied	Transposed fully but late, never applied

necessary for capitalism to succeed, would not take place easily. What, then, can explain differences in the implementation records of different countries? And, at a national level, what can explain the curious partial implementation within individual countries, like Italy and France?

2

The French Case

The Importance of the Policy Legacy

France transposed the Equal Pay Directive fully and on time but, surprisingly, took much longer to apply its major principles. The government claimed in 1975 that its Equal Pay Act (EPA) of 1972 would serve as the official transposition and applicative measure. The EPA included all of the directive's major requirements: equal pay for work of equal value, a broad definition of pay, and the nullification of discriminatory agreements. The act did not include, however, serious provisions for application. Rather than take immediate action, the government waited until 1983 (seven years past the official deadline) to produce more concrete applicative measures. Why was France quick to incorporate legally the principles of the directive but then waited until 1983 to take serious applicative measures?

I argue in this chapter that, at the time of the EPD's enactment, the French state had a unique, deliberate, long-standing approach to the field of female work rhetorically open to wage equality but purposely void of any real administrative control over wages. The demands of the 1975 directive challenged the status quo and asked the state to actively intervene to ensure wage equality between the genders. The French state replied by introducing no change: it used the EPA to transpose adequately the EPD's principles and the same, inadequate, act to apply the directive. To do otherwise—to comply by building the appropriate administrative apparatus for wage controls—would have contradicted years of deliberate nonintervention.

The eventual applicative measures of 1983 represented no sudden decision to comply with EC law and overhaul an existing administrative status quo. Rather, they reflected a gradual shift, taken place between 1975 and 1983, on the part of the state away from conservatism and toward progressive legal and administrative practices. The EPD in 1983 could be applied only because by then it caused no administrative shocks.

The directive's challenges extended beyond the French policy legacy and well into the labor market itself. Women, although often employed in jobs identical or comparable to those of men, in 1975 were earning around 78 percent of men's wages (Eurostat Review 1980): the EPD's principles would have caused a sudden dramatic transfer of resources away from organized working men and management toward women. This uneven distribution of resources was no random historical fact: it reflected decades of union apathy and male domination, and women's related inability to pressure the state to introduce radical changes in the labor market. That inability persisted into the mid-1980s. It left the state free to protect its legacy and the perceived health of the French economy.

Time and Extent of Transposition and Application

In 1965, a time when the EC Council was only contemplating enacting a law that would enforce the principle of equal pay, the French Ministry of Labor was creating a study group that in 1971 became the Committee on Women's Employment (Lorée 1980, 81).[1] That committee produced the drafts of the EPA of 1972, which outlawed discriminatory earning differentials in the private sector, where over 50 percent of women were employed at the time.[2] Upon the enactment of the EPD, the French government informed the European Commission that its 1972 act would serve as its official transposition.

By virtue of already having a clause on equal pay for 'work of equal value' in Title 1, the act could comply with the most important requirement of the EC directive. The act also defined 'pay' broadly, to include base salaries, bonuses, and fringe benefits.

Titles 2 and 3 effectively nullified all discriminatory agreements. Title 2 ordered that the same unit of measurement used to evaluate jobs held traditionally by men and their pay be used to evaluate women's jobs: "the different elements composing renumeration must be established according to identical norms for men and women." These requirements were extended to all types of wage systems in private firms and all the remaining occupations not covered by the existing egalitarian but limited Industrial Relations Labor Code, such as farm laborers (Lorée 1980, 86).[3] Title 3 more explicitly nullified all contracts, collective agreements, documents or previous laws which, openly or in more subtle language, "might transgress the principle of equal pay for work of equal value" (Lorée 1980, 86).

The application of the law was in principle ensured through two major mechanisms.[4] Title 5 reserved the authority to enforce the act to labor inspectors. Inspectors, through routine investigations, were expected to handle complaints and enforce the provisions. Secondly, violators were subject to a criminal penalty in the form of a fine of 600 to 1,000 francs, or approximately 150 to 250 dollars. Employers found guilty a second time within the period of one year faced a fine of 2,000 francs and a ten-day jail term. Employers faced smaller fines if they failed to post the act or to disclose information related to pay differentials to investigators.[5]

As the critics of the bill predicted, these applicative measures were too weak to have a real impact on pay scales. Most investigators were given no specific guidelines for measuring 'work of equal value' and thus limited their search to jobs that appeared practically identical.[6] They found it almost impossible to trace discriminatory practices in the "intricate maze of pay structures" then in existence (Lorée 1980, 88). On the sanctions' side, no time limit for resolution was set, and employers could request counter-investigations. Fines were anyway too small to prevent employers from trying to avoid having substantial increases in production costs.[7]

By September 1975, the Committee on Women's Employment called the act a relative failure.[8] As of February 1975, only eighteen complaints had been registered by the Department of Labor.[9] Twelve of these were soon dismissed. By December 1978, a slightly larger number had reached the courts (Commission of the EC 1979, 44). Only a few cases had been settled out of court (Lorée 1980, 88). The EPD had not been properly applied.

With the Left in power in 1981, a new Ministry of Women's Rights (MDF), headed by Yvette Roudy, was created. The ministry turned immediately to employment and the applications shortcomings of the EPA.[10] The result was the law of 13 July 1983, the 'Roudy Law'.[11] First, the law elucidated the concept of 'work of equal value'. It specified "a restrictive list of jobs reserved for each sex" (Bouillaguet-Bernard and Guavin 1988, 183) and then went on to define 'work of equal value' as:

> Work that requires the same (1) occupational knowledge evidenced by qualifications, a diploma or degrees or professional experience; (2) abilities based on acquired experience on the job; (3) responsibilities; and (4) physical or mental burdens.

Secondly, and most importantly, the law strengthened practical enforcement in a variety of ways. Firms with more than 300 employees (50 by 1986) were required to produce annual reports on the conditions of employment of both male and female workers; the reports had to cover pay levels and then specify the qualifications and classifications used to evaluate workers (Bouillaguet-Bernard and Guavin 1988, 183).[12] In this way the ministry avoided having to collect data. Trade unions and other representative groups were permitted to represent aggrieved women in court; plaintiffs did not even need to be present.[13] Catch-up plans by employers would become part of collective agreements at the firm or sectoral level, thus applying to all women in comparable situations. A network of ministerial delegates would help unions reevaluate the value of work done by women and establish programs to promote equality (Jenson 1988, 163).[14] Thus, inspectorates and employers' reports were not the only mechanisms for identifying discriminatory practices. Their effectiveness was further increased when, soon after the enactment of the law, a new council composed of representatives of employers' organizations, labor representatives and experts was created within the MDF and given the responsibility of disseminating information, receiving annual reports, and developing plans for positive discrimination.[15]

The Roudy Law was well crafted and effective, enjoying "widespread popularity" among different political parties, the public, employers, and even feminists (Stetson 1987, 156).[16] Most employers produced reports and willingly adjusted, in cooperation with unions, pay scales. Most reports show France having the smallest decrease in wage differentials between the late 1960s and 1980 in Europe and the highest decrease in wage differentials between 1980 and 1985.[17] Much research has pointed to Roudy's preemptive negotiations with management and labor as crucial to the success.[18] With the Roudy Law, France finally took the necessary measures to apply the EPD.

Why did France, after transposing the directive effectively, postpone its application until 1983? The following sections investigate this puzzling question.

Legal Progressivism and Administrative Conservatism

The Equal Pay Act of 1972, with its advanced principles of gender equality and its disappointing applicative measures, was deeply rooted in France's social and legal history. The act had emerged as part of a complex, partly progressive but mostly conservative, state effort to regulate the rapidly growing female labor market. Because of the prevalence of conservative influences, the state allowed for the introduction of a law on equal pay but took no serious initiatives to apply it. Whence did the 1972 act come?

Starting in 1962, the postwar trend of a declining employment rate for French women reversed.[19] The number of economically active women increased steadily to become, in 1982, the largest (percentage-wise) in the four major labor markets of Europe. This was the result of at least two factors: improved educational attainment[20] and the need for a second family-income spurred by the industrial crisis of the late 1960s.[21]

Table 2.1 shows the employment of women between 1960 and 1982 compared to that of men, both percentage-wise and in absolute numbers, in France, Britain, Italy, and West Germany. Between 1960 and 1982, Italy witnessed a 5.2 point increase in women as a percentage of the working population, West Germany a 1.7 point increase, and Great Britain a 6.7 point increase. France, on the other hand, saw its female work population increase by 8.1 percentage points. Only Great Britain witnessed a feminization of its

Table 2.1. Number of Working Women (in '000s) and Ratio of Working Women to Working Population in France, Italy, Great Britain, and West Germany, 1960–1982

	1960	*1970*	*1975*	*1980*	*1982*
France	6,611—33.4	7,601—35.5	8,277—36.9*	8,391—38.4	n.a.—41.5*
Italy	6,258—29.5	5,232—26.8	5,563—29.3*	5,915—28.9	n.a.—34.7*
Britain	8,015—32.7	8,937—35.3	9,720—38.8*	10,401—38.9	n.a.—39.4*
W. Germany	9,898—37.3	9,638—35.9	9,818—37.0*	9,953—38.4	n.a.—39.0*

Sources: Data from OCSE, *Demographic Trends 1950–1990* (Paris, 1977); Maria Vittoria Ballestrero et al. (eds.), Lavoro Femminile, *Formazione e Parità Uomo-Donna* (Milan: Franco Angeli Editore, 1983), 189–93.
* Based on data from Eurostat, *Review 1970–1979* (Luxembourg: Statistical Office of the European Communities, 1980); Eurostat, *Employment and Unemployment* (Luxembourg: Statistical Office of the European Communities, 1987).

workforce close to that of France. The new British female workers were headed, due to decades of union underrepresentation and government resistance, to provide the cheap, underpaid labor that somewhat sustained British competitiveness internationally. Rather than regulating the female labor market, the British government facilitated the use of cheap labor from women by adopting a free-market approach. In Italy, trade unions mobilized early on to ensure equality in collective agreements; the legislative arena had proven impermeable since the 1960s.[22] French women could have been used for the same purpose as British women, having entered mostly semiskilled and manual positions,[23] and having, unlike Italian women, little representation in trade unions. In 1975, such a bleak fate seemed precisely to have become reality, as over 50% of the new, lesser-paid, manual laborers were women, accounting for almost 40% of all unskilled manual laborers.[24] French women could have 'sponsored' "the modernization of French mass production industry in the postwar years" and well into the 1970s:

> Women were the semi-skilled and unskilled workers on the assembly-lines, the employees in new mass merchandising, the state workers providing social programs and the second family income on which the whole system came to depend. (Jenson 1988, 158, 159)[25]

The French state, however, reacted forcefully to the feminization of the workforce. Its strategies deserve attentive analysis. They determined not only the fate of thousands of working women in the 1960s, and early 1970s, but also their fate up to 1983: the implementation of the Equal Pay Directive was in fact determined by the early reactions of the French state.

Faced with the question of what its role should be in the face of the large scale transformation of the labor market and the concentration of women in the market's lower echelons, the French government moved in an original and certainly different fashion from the generally antagonistic British and estranged Italian governments. The government *intervened* early on, effectively turning the state into the arena where the problems of 'women and work' would be articulated and a number of solutions attempted.[26]

Strategies conformed to three general types, depending on how government officials decided to interpret the huge influx of women. The state mobilized to protect the traditional family structure, elevate women to the status of equal workers, and use the new labor-power as cyclical, part-time work to attack soaring unemployment rates and to provide industry with much needed flexible and less costly workers. These objectives were certainly contradictory: the first and third were essentially conservative, while the second reflected more progressive thinking. A more coherent, concerted response would only occur in the 1980s. The conservative side produced a number of laws and powerful administrative institutions. The progressive side lacked the same impetus: efforts culminated with the creation of a somewhat marginal committee (the Committee on Women's Employment) and a novel law, the 1972 Equal Pay Act, that lacked serious provisions and administrative support for application.

A part of the interventionist French state, then, began constructing a conservative, effective response to the feminization of the workforce by stressing the link between work and family.[27] Under the center-right presidency of Giscard d'Estaing, a new post of Secretary of State for the Status of Women[28] was created in 1974 and filled by the energetic Françoise Giroud, a well-known conservative editor and writer. In 1976, following Giroud's resignation, Giscard downgraded the Secretary temporarily to the status of delegate but in 1978 retransformed that position as Minister-delegate to the Prime Minister for the Status of Women. Once again, a center right activist, Monique Pelletier, filled that post.[29]

Both Giroud and Pelletier worked to articulate and enforce a view of working women as mothers and wives in need of legal protection. They immediately appointed study committees that bypassed the more progressive Committee on Women's Employment.[30] Their goals were well expressed during a parliamentary debate in 1974:

> [To obtain] security finally for women in permitting them to reconcile their responsibilities as mothers, their occupations and their legitimate desire to play, in the same way as men, an active role in society. (Journal Officiel, *Débats Parlementaires*, Juin 1974, 2495; translated by Stetson 1987, 137)

From this standpoint, the government enacted a long and unique (by European standards) list of powerful protective legislations.[31] A major achievement was the law of 11 July 1975 protecting maternity and pregnancy rights by severely punishing employers who used "pregnancy to hire or refuse to hire, cancel a job contract or transfer employment," or to adjust salaries. It ordered employers to find alternative temporary work for those women whose pregnancy barred them from performing their customary jobs. The family income supplement of 1977 for families with children under three or with three or more children was made considerably higher for families with only one wage-earner and thus made it easy for women to stay at home and cater to their families (Bouillaguet-Bernard and Gauvin 1988, 168). Pelletier, convinced that "family-life and professional life" had to be reconciled by the state "without denying the differences between men and women," proposed yet other measures for creative use of flexible time and early retirement to allow mothers "to drop out of work for a while without losing their permanent status or rights."[32]

In 1977, the government introduced a bill for maternity leaves that entitled mothers to a maximum of two years for child care in six-month segments. The minister of labor himself openly admitted that he understood this to be

> a supplementary step in the definition of protective measures which surrounds maternity of salaried workers—dispositions which . . . were conceived in the double interest of mother and child. (Journal Officiel, *Débats Parlementaires*, 16 June 1977, 2584; translated by Stetson 1987, 138)

This time, moderates and feminists spoke up. The chairman of the National Assembly Committee on Social, Cultural, and Family Affairs argued that the benefit should extend to fathers as well. The government disagreed with this argument, but did agree to redefine the bill as 'parental leave' before submitting it to the Senate. Feminists and moderates agreed that, regardless of its official scope, mothers would use the law; in Lane's words, "in France, the state has made it financially worthwhile for women to stay at home with their children" (1993, 279).

While these conservative and protective responses to the feminization of the workforce dominated legislative and administrative activity during the 1970s, a more progressive wave was under formation. The roots of this progressive movement dated back to the postwar years, when the government issued the 1946 General Statute of the Civil Service, amended in 1959. The statute immediately applied the equal pay principle to the entire public sector, "since salaries were based on an index classification and grade rating for each occupation" (Lorée 1980, 83), and affected increasingly more women.[33]

In 1965, following the vision of a senator and member of the French United Nations delegation, Marcelle Devaud, who wished to promote "women's employment rights from within the government" (Stetson 1987, 142), the government assigned an agency exclusively to women's issues: the Committee on Women's Employment (CTF),[34] located in the Ministry of Labor. The CTF was advisory, responsible for analyzing women's employment and training and for submitting policy-proposals for new legislations and rules. Its membership included representatives of unions, employers and women's organizations, including feminists. The CTF's explicit mission was equality and the right of women to work. In its approach, it contrasted sharply with the later study groups, mentioned above, appointed by Giroud and Pelletier.[35] In its 1971 report, for instance, the committee recommended that the government enact a new statute for equal pay with serious sanctions, new policies for job training, and new rules for labor allocation. In a climate of conservative interventionism, the CTF encountered resistance. Of all its major proposals, only the equality of pay one survived the resistance of the Pompidou-Messmer government.

The bill for the 1972 act was drafted by the minister of social affairs, Joseph Fontanet, and was deliberately deprived, soon after its first version, of major provisions that would guarantee a strong practical impact on wage structures.[36] The National Assembly debated and approved the EPA in one sitting on 21 November 1972. The Senate also debated and approved it in one day, 13 December 1972. All amendments proposed in both houses were rejected by a strong opposition[37] and were not re-introduced under a different guise. Most of the proposed amendments would have introduced more severe punishment for discriminating employers and/or more objective and reliable methods for identifying infractions.[38] Amendments were rejected for fears that increasing female labor costs would increase productivity costs and put French employers at a disadvantage with the rest of Europe, and in the explicit belief that women were to function in society as secondary wage earners and therefore did not warrant a lowering of men's wages (Lorée 1980, 85).[39]

The act was judged at the time as a major conceptual breakthrough that could affect over 35% of the total workforce, and 75% of the increase in labor force between 1968 and 1975.[40] Its provisions had failed, however, to build the administrative structures necessary to turn conceptual parity into reality. Specifically, it had not provided for a commission or committee charged with overseeing implementation in a systematic fashion leaving that task to unprepared inspectors. Sanctions, perhaps as importantly, were too small. The architect himself, Minister Fontanet, admitted that the "efficacy of the [bill was] not certain," but then voiced what was the prevalent conservative mood of the time:

A famous Greek civilization historian said that the excess of repressive legislations was either one of the causes or one of the signs of Hellenic decadence. (Journal Officiel, *Débats Parlementaires*, 21 November 1972, 5562; translated by Stetson 1987, 143)[41]

The 1972 act was the work of an assiduous committee. Because it ran against the predominant conservative mood of the times, its applicative shortcomings proved impossible to remove.

There was a second, albeit minor, product of this progressive wave. This was a substantial boost of the minimum wage in 1968.[40] After 1972, the state increased the wage at rates higher than inflation. This was part of an openly admitted strategy to benefit women, who composed a large percentage of workers in the lower-paid industries, such as hairdressing, clothing, and catering (Bouillaguet-Bernard and Gauvin 1988, 171).

The third, more conservative, response of the state to the feminization of the workforce built another, fairly limited, strand of laws. This time the state conceptualized the new working women as cyclical, part-time workers (Jenson 1988, 167; Lorée 1980, 99–100).[43] Women would provide deregulated, cyclical work; men would comprise the stable workforce. The partial use of female labor would accomodate more women and would halt soaring unemployment rates: between 1970 and 1978, female unemployment was twice the male rate, accounting for 60% to 75% of all unemployment.[44]

The Act of 27 December 1973 and the Act of 9 June 1975 thus extended social security and other benefits to part-time work, while making it easier for employers to create part-time positions by, in part, downgrading full-time positions (Lorée 1980, 99). The law of 28 January 1981 permitted employers to institute part-time work at will (Maruani 1984, 135). Women, it was thought, would then naturally fill the new positions: after all, they had held almost 100 percent of part-time positions in the past.[45] To help the natural course of events, the state proceeded to pressure those corporations and businesses employing more women than men to take advantage of the new provisions.[46] Subsidies to all complying companies were then distributed (Jenson 1988, 166). The state's strategy of encouraging especially young, married women to work through protective laws was successful.[47]

As a response to the feminization of the workforce, the French state had enacted *laws* and built an *administrative apparatus* that were primarily conservative, although they included the progressive Equal Pay Act of 1972 and the CTF of 1965. When, in 1981, Mitterrand became president, the more progressive strand of the French policy began to gain the upper hand. Only then did the government move to create the administrative structures capable of applying the principles of equality found in the 1972 EPA.

The Evolution of a Legacy: Administrative Changes after 1975

Mitterrand's rise to the pinnacle of French power altered the conservative approach to female labor in the state. Mitterrand worked with existing laws and administrative entities to produce a more progressive administrative apparatus. He capitalized on the existence of progressive entities, such as the 1972 Act and the CTF to improve the application of the EPA.[48] He cleverly worked to marginalize conservative institutions. In the space of a few years he revitalized the state's egalitarian approach to women's work. With one major law in place, the Roudy Law of 1983, the EPA, and thus the EPD, could finally be applied.

In his first year, Mitterrand introduced three important administrative changes. First, he elevated the Minister-delegate to the Prime Minister for the Status of Women to a Ministry of Women's Rights (MDF).[49] Still legally a delegate to the Office of the Prime Minister, the MDF was granted a separate budget with a tenfold increase in funds, and was committed primarily to employment equality rather than family. Secondly, Mitterrand *moved* the old bastion of employment equality (the CTF) into the MDF. In so doing, he created a vacuum in the area of workers' family protection traditionally handled by the predecessor of the MDF. To solve this, Mitterrand appointed a separate secretariat for family and population within the Ministry of National Solidarity and Social Affairs, itself also active with immigrant workers. This transformation of institutions, only possible on the bases of preexisting institutions, worked to prioritize and focus resources on employment equality and to relegate family protection to a secondary position, in the hands of a ministry busy with additional concerns. In 1985, the MDF was finally upgraded to full ministry with cabinet status (Stetson 1987, 20).

Yvette Roudy, a feminist, was named minister of women's rights. This was an indication of the fact that, through Mitterrand, a number of prominent feminists were gaining access to the legislative and executive branches and that, gradually, the political arena would become the preferred battleground for feminists. Roudy's institutional position demanded that "employment [be] her number-one priority" (Stetson 1987, 146). She used the communication channels between state and social actors, established by the former in the 1970s during its protectionist streak, to fight the conservatives' notions that a woman's salary should be considered secondary income and that a woman be encouraged to work less and fulfill her domestic role. A long period of consultations with labor unions, employers, women's associations, feminists, and other parts of government began.[50] Roudy enjoyed a position of legitimacy both outside and within the government: no one questioned that the state would continue to be involved in the regulation of the labor market of women.

The consultations served to identify the most serious problems afflicting the application of the law as well as to prepare employers to accept the gradual increases that would come from wage equalization. Employers, for the most part accustomed to state intervention, collaborated. They pointed out, for instance, that the blurry definition of equal pay had precluded any possibility of its application. After one and a half years, a bill that enjoyed the support of both labor and managers was presented to the National Assembly (Stetson 1987, 146–48). Conservatives opposed the bill's redefinition of work, since it made no mention of women's family roles. Missoffe, for instance, "warned that the government must recognize that women have different lives" (Stetson 1987, 146). Conservatives were in a minority and the bill passed easily in the National Assembly. It faced some revisions in the less Socialist Senate and became law on 13 July 1983.

The state's creation of an additional administrative unit, dedicated specifically to equality, was as important to the law's success as the recently won cooperation of unions and management. The state created within the MDF the Council for Professional Equality.[51] This replaced the old Committee on Women's Employment and had one major objective: to continue to work with the rest of government and representatives of key interest groups to ensure the application of the new law.[52] Thirty-six members composed the council: these included labor market experts and representatives from all employers organizations (Stetson 1987, 148). They contacted employers and unions, disseminated information, and received the reports that employers were required to produce annually. The average employer was now facing a united, well-crafted government machinery that expected his cooperation.

The interventionism of the French state, and its complex response to the feminization of the workforce, had created a legacy of administrative structures and legal initiatives that, under the influence of Mitterrand, turned toward equality and away from protectionism. The ground for the proper application of the EPD was finally laid.[53]

Working Women and Trade Unions

Working women, subject to significant wage discriminations—they earned 78 percent of men's wages for identical work or work of equal value in 1975 (Eurostat Review 1980)—would have greatly benefitted from a prompt application of the EPD. Why, then, did they leave the state free to dodge the EPD's principles? Why did they not use the weight of a European Community directive to overhaul a conservative legacy and advance their position in the labor market?

Women had unions to rely on for support. These, however, had repeatedly proven to be weak organizations with no real interest in advancing gender equality. This is why the French state had been free in the first instance to appropriate onto itself the right to regulate the feminization of the workforce in the 1960s and 1970s. There were no organizational grounds for women to mobilize in 1975 and later and ensure the introduction of a radical directive in the labor market.

Unions were in the first place weak organizations. One measure of the unions' general weakness was given by their density: union membership as a percentage of all employees. Table 2.2 shows French unions and their density along with figures from the other four major industrial powers. In 1970, the early period of state intervention and the heart of the feminization process of the workforce, French unions were weaker than unions in all major economies.[54] In the critical period 1970–77, French unions were the only ones to show weakening; in all other countries of the sample, unions enjoyed a period of expansion.

Other measures of weakness confirm the figures in table 2.2. French unions enjoyed, during this period, fewer institutionalized channels for bargaining, used more confrontational strategies, espoused utopian visions, and, most importantly, settled fewer collective agreements than most European unions (Kesselman 1984, 2). "Unable to compel employers to bargain collectively over wages, hours, and working conditions," Kesselman wrote, organized French workers could not "achieve comparable benefits to workers elsewhere" (1984, 2). In France, the state produced and "had the power to make discretionary increases in the minimum wage" (Bouillaguet-Bernard and Gauvin 1988, 169–70); in Italy, Germany, Denmark, Belgium, and the Netherlands unions did that (Rubery and Fagan 1993, table 1.4, 12, 34; Lane 1993, 290).[55]

Organizational weakness aside, unions would not function as a weapon for gender equality for a second, perhaps more important reason: ideologically, unions preoccupied themselves with the traditional topics of male workers' wages and benefits. Women, as well as immigrant workers, environmental protection, organization of work, and technological change remained marginal topics (Kesselman 1984, 3).[56] Both the Confédération Française Démocratique du Travail (CFDT) and Confédération Générale du Travail (CGT), the two largest unions in France, developed only theoretical debates during the 1970s concerning working women. They had never mobilized, either for (as was the case for the Italians) or overtly against (as was the case for the British) equality in the spheres of collective agreements or politics.[57]

Of the two unions, the CFDT alone showed some inclination toward women's equality.[58] Its new radical vision of *socialism démocratique et autogestionnaire* made the economic liberation of women an important step for the construction of socialism. The exploitation of women was contextualized within the larger mode of capitalist and male-dominant production. For that reason, according to the working documents of the 1970 congress, the

Table 2.2. Union Density* in 1970 and Changes in Density (Δ) during 1970–1979 and 1980–1989 in Five Countries

	France	United Kingdom	Italy	West Germany	Denmark
1970	21.5	48.5	37.5	37.7	61.7
1970–79 (Δ)	−2.9	5.90	16.0	3.90	17.8
1979	18.6	54.4	53.5	41.6	79.5
1980–89 (Δ)	−7.4	−11.2†	.100	−3.1	.100‡
1989	11.2	43.2	53.6	38.5	79.6

Source: Data from Bruce Western, "A Comparative Study of Working-Class Organization." *American Sociological Review* 60.2 (April 1995).
*Density: union membership as a percentage of all employees.
† 1980–1987; ‡ 1980–1988.

liberation of women would occur both at the workplace and ideologically in civil society. At the work place, the CFDT identified women's poor working conditions, *lower salaries*, and vulnerability to unemployment as serious problems deserving of attention (McBride 1985, 50). In civil society, the CFDT identified education, reproductive policies, child care, and even union structures as equally important problems (Jenson 1984, 164).

Practice did not follow ideology, however, at least in the crucial decade of the 1970s. The CFDT "was not very explicit in its national-level documents about how the project would unfold" (Jenson 1984, 164). When pressed, the CFDT refused to develop a program for specific reforms for working women. In 1971, it abolished its women-only Commission Confédérale Féminine, replacing it with a mixed body (Jenson 1984, 165). Throughout, it consistently opposed the creation of a separate printing press for female unionists. There was never a moment when the CFDT contemplated a large-scale mobilization for equality of wages, as pressuring the state for the application of the EPD would have required. Indeed, in the 1980s, officials began recognizing the "CFDT's failure to live up in concrete ways to the promise of its theoretical analysis" (Jenson 1984, 173). Belatedly, federations and local unions were encouraged to institute separate women's commissions, conferences on Women's Work and Union Action were held, and women were put in positions of leadership to fill a 25 percent quota.[59] By that time, the Socialists in government were already capitalizing on a large administrative structure to ensure the principle of equality.

The CGT conceptualized the problems of working women more conservatively, centering its rhetoric on the particular needs of those who were married. In its 1973 National Conference of Salaried Women report, the "conceptual link between 'women' and 'mothers' remained very close" (Jenson 1984, 162); the CGT openly opposed what it called "integral egalitarianism" (Maruani 1984, 136).[60] It defended its position by claiming that it understood the 'real situation' of working women, and by accusing those with more egalitarian aspirations of utopian idealism.[61] The "increased participation of women in wage labor" and the struggles that this caused, argued a female member in 1973, promoted "a better conception of married life and the division of responsibilities within the family" (quoted in Jenson 1984, 163). Accordingly, the CGT advocated "women's retirement age . . . set at 55, with additional benefits for each child raised" (Jenson 1984, 162), maternity leaves, and day-care facilities (Stetson 1987, 132). It attacked its most progressive members, such as the editors of the magazine *Antoinette*, and those who supported progressive ideals, like equality of pay.

The CGT hardly mobilized in pursuit of women's issues anyway. Its theoretical debates were in fact replicas of the French Communist Party's ideas (an institution on which the CGT's survival depended), and were intended to show conformity with the Party's ideas rather than true, action-oriented commitment.[62] As the Communist Party's position on gender issues changed,[63] the CGT merely followed without ever feeling obliged to translate the new directives into real social change: the CGT was acting to "best match its own union strategy to the needs of the [Communist Party]" (Jenson 1984, 163).

Constrained by resources and ideological commitments, both the CGT and the CFDT played a minor role in the regulation of the female labor market. This, aside from leaving ample room for the state to intervene in the increasingly 'feminized' labor market of France, left women with no real representation in the marketplace. The Equal Pay Act of 1972, with its limited applicative measures, was the clearest indication that the state, and not the unions, determined how female work would be regulated in France. Where would women, only three years later, find the resources to pressure the state to take on a radical departure from past practices and introduce an unprecedented, highly transformative directive in the French labor market? Unions were certainly not that place.

Explaining Implementation

The policy legacy of the French state explains the implementation of the Equal Pay Directive in France. Legally speaking, the directive asked the French government to transpose principles already present in its own national legislation. The Equal Pay Act of 1972 was fully established in the French legal system and included all of the major principles of the directive. By pointing to the 1972 act, the French government could safely state to the European Commission that it had complied with the requirements imposed upon the country by the directive. Transposing the directive by relying on the 1972 act posed no challenges to the legal tradition of the country.[64]

Truly accounting for transposition in France, then, means in a sense to recognize the presence of the 1972 act and the circumstances that brought it about. As seen, three crucial variables produced the 1972 Act: the rapid feminization of the workplace, the resulting state's decision to intervene in the labor market, and a general silence on the part of trade unions that made that intervention possible. Had there been no feminization of the workplace, or no intervention by the state, or more active (progressively or conservatively)[65] unions, the 1972 act might not have existed. The directive would have then undergone, potentially, a very different transposition process.

Full application of the principles of the directive proved far more problematic. The French state would have had to apply the 1972 act. There were, however, deeply rooted reasons for the lack of administrative support and sanctions that had neutralized the impact of the 1972 act. French legislators had already proven themselves against the application of the principle of equal pay three years prior to 1975: despite isolated objections (such as those of the Socialist Michel Rocard) the 1972 act had consciously been stripped of any serious applicative measure. *By pointing*, in 1975, to the same act it had made impossible to apply, the French state thus ensured that nothing emanating from the EC would alter its current stance toward the regulation of female labor.

The intervention of the French state in the labor market took the form of institutional building. Despite major conservative influences, a Committee on Women's Employment, a Secretary and a Minister-delegate to the Prime Minister for the Status of Women were instituted. Application had to wait until a natural evolution of the French policy legacy revitalized the progressive principles that existed in the state's response to the feminization of the workforce. The evolution, brought about by Mitterrand, was possible only by a rearrangement of existing administrative and legal entities. Once the French state had redirected its own intervention in the labor market, the directive could be applied.

The organization of two interest groups was challenged in 1975 and in 1983. In 1975, women would have benefitted from the directive but could not mobilize to ensure the application of the beneficial EPD. Throughout the post-1983 period, employers, as the interest group most negatively affected by the directive, should have opposed the application of the directive. The state was able to preempt a likely negative reaction of employers by using long-established channels of communication and cooperation and by creating special conditions, such as the Ministry of Women's Rights and the Council for Professional Equality, for consultations and collaboration. It was ultimately the presence of a favorable French policy legacy, and its natural evolution through time, that determined the fate of the EPD in France.

3

The Italian Case

Strong Union Representation

To the surprise of those skeptical of the speed of the Italian legal system, Italy transposed the Equal Pay Directive fully with only one year of delay. After an early attempt to argue that its Constitution's commitment to equal rights made further legislation unnecessary, Italy passed Law 903 of December 1977. The law satisfied all the major principles of the directive. Just as in France, however, its application was weak. The law provided for no concrete measures to ensure information flow, assistance to applicants, enforcement, and penalties.

In this chapter I account for the directive's implementation by looking at its demands on the Italian policy legacy and the organization of Italian interest groups. The directive was well transposed because, thanks to advanced collective agreements in place since the 1960s, it implied little change for the resources of most working women or the resources of employers. Legislators found it easy to draft quickly a noncontroversial, even if unusual, law. The EPD was not well applied for two reasons: because it asked the state to intervene in a new field where it had no administrative capacity; and, less importantly perhaps, because it challenged the organizational bases of some women who, in the 1970s, had turned away from ideologies of equality.

Time and Extent of Transposition and Application

In 1975, the Constitution guaranteed equality of pay for identical work for women in the labor market. Article 37 provided that "a working woman shall have the same rights and shall receive the same pay for equal work as a working man." Law 533 of 11 August 1973 made the application of Article 7 easier: taking an employer to court became simpler and less costly, with the state absorbing some of the costs. Equal pay for work of equal value was practically guaranteed by several broad interpretations of Article 37 by the courts in favor of claims concerning 'work of equal value' and, especially, by collective agreements at the national level signed in the early 1960s between unions and employers. In light of this and with one of Europe's lowest gender wage differentials, Italy claimed in 1975 that it fully complied with the Equal Pay Directive (Landau 1985, 77). The European Commission rejected the claim and insisted that a legislation on 'work of equal value' be drafted (Mazey 1988). Italy proceeded without any delay.

At the opening of the VII Legislature, in 1976, thirteen drafts for the new law were submitted, eleven of which came from newly elected women in Parliament. Two drafts came from the Independent Left and the Socialist Party; four from the Christian Democrats; five from the Communist Party; two from the Social Democratic Party; and even one from the Movimento Sociale Italiano, the neo-Fascist Party. The discussions took unusually little time, and a single draft of nineteen articles was unanimously approved by Parliament.[1]

The most relevant aspect of Law 903 was Article 2. It provided that "female employees are entitled to receive the same pay as male employees for identical work or work of equal value" and that "the systems of job classification used to determine pay shall adopt common criteria for men and women." There was, importantly, no mention at all of 'material defense' exceptions as in the British revised version of the EPA in 1983, examined in the next chapter. In fact, Article 2 specified that classifications were prohibited when implicitly organized around gender. Classifications which ranked jobs traditionally occupied by males as higher needed thereafter special justification (Ballestrero 1983).

The law defined pay broadly, in accordance with the directive. Pay included, in Articles 8 and 9, coverage for injuries at work, illness, pension schemes, and benefits to family members. Article 4 allowed different retirement ages for males and females: women were now permitted, not obligated, to retire earlier than men with equal pension benefits.[2]

In conformity with the EPD, Article 19 voided as illegal all collective and individual contracts, and international regulations of firms and staff that were contrary to its rules. It thus nullified, although the EPD did not directly demand this, all protective laws, such as Law 653 of 1934, that prohibited women from working at nocturnal, dangerous, unhealthy, or physically demanding jobs.[3] It demanded that no discrimination be made in any sphere that was directly or indirectly determinant of pay. Article 1 prohibited discrimination in hiring, training, and periodical retraining; Article 3 prohibited discrimination in career advancement structures.

As such, Law 903 faithfully transposed the directive into national law. Criticisms against the law were certainly raised at the time and later, but these applied to the content of the directive itself and not to its translation into Italian law. Critics for instance argued that the directive, and consequently Law 903, focused on discrimination between subjects that were equally qualified but received different wages solely for gender reasons. Inequality of opportunity, they insisted, was being ignored.[4] The unusual agreement on the content of Law 903 among different societal representatives was noted by a major union's magazine:

> During the approval procedure of the law the unanimity among the parties, feminist organizations, unions and experts is broken only by a few isolated voices. (*Rassegna Sindacale*, 21 June 1979)

The application of Law 903, by contrast, was poor. In the opinion of many critics, Law 903 offered "the weakest instruments to eliminate discrimination . . . it [was] a symbolic measure, consciously neglectful of implementation problems" (Beccalli 1985, 450). The most optimistic and naive supporters realized almost immediately after its approval that the law would have no impact (Ballestrero 1983, 9).

The burden of proof fell on the woman. No guidelines for job classifications were prescribed, making judges in tribunals fully responsible for evaluating employers' criteria. Sanctions were minimal. Article 16 provided that "any person failing to comply with the provisions contained . . . in Article 2 . . . shall be liable to a fine of between 200,000 and 1 million lire," approximately 1975 150 to 700 dollars. There was no retroactive compensation, and, importantly, cases *applied to single women* only and could not extend to other women in similar conditions.

As trade unions argued, the idea that implicit gender biases in job classifications could actually be spotted by judges was unrealistic. Occupations with a high concentration of women, as in the textile industry, had traditionally lower wages without direct or indirect mention to gender as a criterion of evaluation (Ballestrero 1983, 26). Many pointed out how "job grading structures reflect[ed] deep historical practices and social structures and social attitudes toward particular forms of work" (Rubery and Fagan 1993, xviii) that would be impossible to catch.

Trade unions were charged with ensuring that collective and individual agreements offered equal pay.[5] The government's direct involvement in the application of the law was thus avoided. Only Article 18 stated that: "Government is expected to submit to Parliament a report on the application of the law." This would include the number of cases, encountered problems, the opinions of trade unions, and other relevant information. But there was no mention of any committee or administrative structure charged with any aspect of implementation. Legislators seemed sure that collective agreements already in existence could, as they had already anyway, ensure equality.[6] Trade unions, feminists, leftist parties' parliamentary members posed, surprisingly at first sight, no objections to such optimistic attitudes.

In 1982, two 'ghost committees' were instituted by the minister of labor to oversee the implementation of the law. Their supposed functions included information dissemination to workers, legal support to applicants, and research activities. These committees never came into existence. A ministerial decree created the Committee for the Attainment of Equality on 12 June 1984. Before drowning in bureaucratic complications, this committee consulted, submitted proposals to the minister of labor, disseminated information and funded research projects. In 1988 and in 1991, with Law 125 of 1991, more powerful commissions were created. Their scope was much broader than equality of pay since by that time women were focusing on equal opportunity. By 1991, Law 903 was a forgotten legislative piece.

Empirical data show that Law 903 had almost no impact on wage structures and differentials (Ministero del Lavoro e della Previdenza Sociale 1985; Ballestrero 1983, 25).[7] A report of the Ministry of Labor in 1985 on the effects of application states that "since its approval, the law has had an almost nonexistent impact" (Ministero del Lavoro e della Previdenza Sociale 1985).

Law 903 faithfully translated the Equal Pay Directive. Italy was among the first to properly transpose the directive in Europe. The application of the law was, surprisingly, almost nonexistent. The following sections investigate the causes of this puzzle, starting with an analysis of why transposition was so expedient.

Strong Organization: The Favorable Collective Agreements of the 1960s

The Equal Pay Directive introduced ideas about equality already present in the Italian labor market that had been introduced and applied primarily through collective agreements, following a long history of favorable union representation. Drafting a faithful translation proved easy; ensuring its application required commitments and administrative institutional building that the state was unwilling to provide. By examining the history of women's labor representation through unions up to 1977, this section accounts for the preexistence, dating to the early 1960s, of the principle of equal pay for work of equal value in Italy's collective agreements and the consequent ease with which legal translation took place.[8]

The relationship between trade unions and women underwent three phases: 1890–1914, 1914–45, 1945–77. During the first phase unions were born, for structural and ideological reasons, favoring female membership and female equality within their ranks and in the labor market. World War I and Fascism put a temporary stop to all union activities. From 1945 on, unions, true to their past commitments, actively pursued equality. They guaranteed 'equal pay for equal work' in the legal sphere in 1947 and 'equal pay for work of equal value' in all collective agreements between 1960 and 1963. In the remaining years, they focused increasingly on alternative objectives. Table 3.1 outlines these events.

Between 1890 and 1914, Italy's late but rapid industrialization began, along with the formation of a heterogeneous unskilled working class and new trade unions that replaced the few existing craft unions of the previous era.[9] Unions were of two types, Socialist and Catholic, organized respectively under the Socialist-Communist oriented Chambers of Labor and the Christian Democratic Labor Unions.[10] Being new, unions *competed* heavily for members. The "overriding aim of both sides," writes La Vigna, "was to win the greatest number of workers in their respective unions" (1985, 129).

Concentrated in the largest economic enterprises of the country, agriculture and the textile industry, where they made up 40% to 75% of the workforce, women became the beneficiaries of this warfare. Unions used newspapers, *Il Lavoro* for the Catholic and *La Difesa delle Lavoratrici* for the Socialist, to attack each other and urge women to join the ranks. Consistently, between 1890 and 1910, articles and advertisements in these newspapers guaranteed improvements in wages and working conditions in exchange for joining.

Ideology fueled, and was partly the basis of, competition. The Socialists in the Chambers of Labor thought that the formation of a united working class was only possible with a homogeneous collectivity. Working women, as part of the working class, should therefore be recruited into unions and turned into conscious class members. Only in this way could the class consciousness essential to the new society of the future be formed. In an article, *La Difesa* stated: "Ours is a class movement that we salute as the indication of the political and economic awakening of female proletariats" (6 April 1913, 1). The Socialists attacked the Catholic unions for forming a "dam against the free proletariat movement" by drawing away workers from the class struggle. They warned their "female comrades . . . about the tentacles of the Jesuitical priests" (17 March 1912, 1), who interpreted the Gospel as they wished and dissuaded the workers from rebelling (19 May 1912, 2). Such unions were too weak to deliver on their promises: "clerical unions cannot be anything but scabs: it is in their nature" to be unable to sustain a strike (22 September 1912, 1), for they are actually partners of the middle class, and not of the true workers (6 April 1913, 1). They urged all working women to join the Socialist ranks for the fight against their conditions, for the liberation of all working women (6 April 1913, 1), and asked them to assume an "anticlerical" attitude (18 January 1914, 1).

Catholic unions depended heavily on the presence of women for survival. Framing their fears in terms of a war against the "atheist Socialists in the

Table 3.1. Women's Strong Representation in the Italian Labor Market, 1890–1977

	Female Union Membership	*Leadership/Groups within Unions*	*Major Policies Pursued by Unions*
1890–1905	Low in absolute, high relative to Europe	Altobelli head of largest union, internal commissions	Active recruitment of women, equality of pay, protective legislation
1905–1920	"	All women groups, esp. in CGL	Equality in some collective agreements, protective legislations
1920–1945	Almost none	Almost none	Very few, but included equality
1946–1969	High again	Several commissions and groups	Equality in Constitution, coll. agreements ('60–'63) ensure equality for work of equal value in all sectors
1970–1983	"	Collectives	Under 'new' feminism, pursuit of policies other than equality

Chambers of Labor," they answered in kind (La Vigna 1985, 129). They accused the chambers of hypocrisy, of instrumentalizing women for political goals that had little to do with the interests of women and much with that of the Socialist Party, of which, they claimed, the chamber was simply a "section."[11]

Lost as subjects of debates, women at times did not see any real action in their favor.[12] The commitment was genuine however, and some advances were made. The Chambers of Labor, following a formal declaration in favor of equal pay in the *Statute of Milan of 1891*, organized women's commissions in all national territories. In 1905, a woman, Argentina Altobelli, became the national leader of the agricultural union, the largest in the country. Between 1900 and 1914, a number of smaller women's organizations, such as the Unione Femminile Socialista, were set up under the auspices of the Socialist Party (PSI) or the Catholic party and their unions to investigate specific female workplace issues. *La Difesa*, the PSI union's newspaper, urged women to hold meetings and courses on culture for the intellectual elevation of women in order to understand better their rights and know how to mobilize (*La Difesa*, 3 March 1912, 3). Protective laws and regulations, like regulation No. 818 of

November 1907[13] on nocturnal, underground, unhealthy jobs, were the primary, concrete victories of this period.

The Socialists intensified their commitment toward equality in the years prior to World War I. Following the reorganization of their unions into the Confederazione Generale del Lavoro (CGL), at their first Congress of 1911 in Padua they launched a major drive to recruit women. The objective was, again, to convince "female comrades at work to join trade unions," to free themselves from the traditional family and social roles which inhibited their pursuit of freedom. "Every worker," wrote a leading exponent in *La Difesa*,

> must persuade the women in his family to fulfill their elemental duty . . . and when we will have crossed the threshold into class organization, we will be able to count on a potent force and shall be able to attempt to good effect the rewarding task of working class redemption. (7 January 1912, 1)

Wage equality was the first priority: Kuliscioff, a leading woman figure in trade unions, wrote in that issue that "strong through its solidarity, the trade union will then be able to raise salaries and shorten hours of work. From involuntary scabs against men, women will become their blood and faithful allies."[14] Two women, Carlotta Clerici and Altobelli, were awarded seats in the National Council of Labor. Then, a crusade was started to convince all male workers to support their wives and children. An issue of *La Difesa* was included in every issue of the *Avanti!*, the party's popular newspaper, for that specific purpose. Some wage increases in the textile sector were obtained in negotiated contracts.[15]

The start of World War I and the subsequent Fascist period precluded the organization of major CGL wage campaigns. By 1914, however, a unique relationship between unions and women had already been established. The resulting low levels of discrimination, and the unusual ability "to integrate the feminism of the time" into their ranks would later push unions to embrace the cause of women and equality of pay (Beccalli 1985, 427).

During the Fascist period, in fact, unions could not achieve much. In 1921, the year Mussolini came to power, Fascist squads burnt local Socialist and trade unionist headquarters and used violence to intimidate trade union members. The state organized the subordination of women, following the slogan 'devoted spouse and exemplary mother', in the *fasci femminili*. A series of laws prevented women from competing for or holding positions as managers, school directors, administrators, and so on (Galoppini 1980, 109).[16] Another series of laws sought to protect women from work; longer maternity leaves, for example, were granted.[17] The state found ideological support from the papal encyclical *Casti Connubi* of 1931, Pope Pio XI's statement on woman as wife and mother. Early on, however, women took part in the Resistance and joined outlawed unions, whose commitment to equality remained unshaken, as proven at the end of the war. Unions encouraged women to form alternative, lawful groups, which eventually merged to form the Unione Donne Italiane.

Immediately after 1945, unions mobilized to fulfill their early promises of equality of pay, an effort uniquely Italian in the Europe of the time (Beccalli 1985, 429). New women's committees and internal groups were organized within the Confederazione Italiana Generale del Lavoro (CGIL), the only union at the time, created in 1944 with the participation of the Socialists and the Christian Democrats. Mobilization occurred on two fronts: collective agreements and legislations.

On the legislative front, achievements were obtained primarily in the immediate postwar years. Efforts were concentrated on influencing the new Constitution in favor of equality. Unions had a particularly easy access to the Assembly charged with drafting the document for two reasons. The PCI was, along with the Christian Democrats (DC), the creator of the Constitution: the historical ties to the party meant that those party members involved in the drafting of the document were either sympathetic toward the unions or representatives of such unions (Sassoon 1986, 15–30; Beccalli 1985, 426n2).[18] Moreover, an unusual percentage of the members of the Assembly were women with older ties to unions and feminist ideology.[19]

With union representatives and sympathetic party members, such as Teresa Noce,[20] equality of pay was introduced and accepted, without opposition, in Article 37 of the new Constitution (Galoppini 1980, 154). The article provided, as the British Equal Pay Act of 1970 did much later, that "a working woman shall have the same rights and shall receive the same pay for equal work as a working man." The subsequent interpretations given by the courts, the government, and trade unions were very broad, although this did not satisfy the EC Commission in 1975: all consistently took 'equal' to mean both "like work and work of equal value" (Commission of the EC 1979, 11).

The affiliation, not found in Britain, of trade unions with political parties granted unionists a number of seats in Parliament from the first legislature until 1969. This was a peculiar aspect of the Italian state, one that reflected again, and strengthened, the historical ties between parties and unions.[21] The presence of unionists was far from negligible. They were 10 percent of all parliamentary representatives in the early 1950s and 1960s (Centro di Studi Sociali e Sindacali 1984). Their actual weight increased when parties offered unanimous support for affiliated unions.[22]

Unionists in Parliament fought for the rights of women. In the first legislature, 28.4% of all law proposals submitted by unions concerned solely women; in the second, 18.7% of proposals concerned women; in the third, 9%; and in the fourth, 5.1%. At least five attempts on equal pay were made in the second legislature.

The success rate was minimal, however; this may explain the decrease in effort over the years and the shift of efforts to collective agreements. Only two proposals became law. In 1950 and 1963, laws on unfair dismissals and protection of working mothers were obtained with the help of Communists Teresa Noce, Venni, and Venegoni (Ballestrero 1979).[23] A second victory occurred in 1956 when unions obtained the government's ratification of the International Labour Organisation's 1954 convention on equal pay for work of equal value, placing Italy among the most progressive countries in gender legislation in Europe. After 1970, unions obtained Law 533 of 11 August 1973, which facilitated the application of Article 37 of the Constitution by asking the state to absorb some of the plaintiff's costs.

As opposed to France, collective agreements proved a more favorable forum for debate. Days after partisans freed regions from German occupation, unions fought for and won equality of pay in Northern Italy for textile and wool workers (Galoppini 1980, 154). With this began a period of *"explicit"* commitment and work toward pay equality that would reach its climax in 1963 (Beccalli 1985, 432, emphasis added). In 1945, the CGIL organized a congress for the new postwar Italy where it claimed to be in favor of equal opportunity for women. Between 1945 and 1947 aggressive defense of the lowest-paid workers led to an increase in wages mostly for women and thus decreased wage differentials between the genders. In 1947, at the first postwar national Congress, the CGIL reemphasized its commitment by releasing an official motion on the subject, "Mozione delle Lavoratrici Italiane" (Beccalli 1985, 431). That same year, it won equality of pay in agreements for some sectors of industry and decreased differentials by half in other sectors.

In 1954 a monumental national conference on equality was staged in Florence by the CGIL.[24] 20,000 firm-level meetings were held before the conference, which one thousand female union officers attended. It was followed by debates, mass meetings, other conferences, firm-level assemblies, and the production of evidence on wage discrimination. In accordance with its resolutions, the CGIL immediately began talks with industry leaders. By that time, it was representing more than 70 percent of all union members.[25]

In 1957, after splitting from the Confederazione Italiana dei Sindacati dei Lavoratori (CISL) and the Unione Italiana del Lavoro (UIL), the CGIL managed to guarantee equality of pay, for its members, in all private sector collective agreements. Then, in 1960, the CGIL, CISL, and UIL obtained the elimination of all discriminatory classifications based on gender in the manual industrial sector and introduced objective codes for evaluation at the national level (Beccalli 1985, 433–34). The agreement took place on 16 July 1960 between the three unions and management after "long and tiresome meetings" (Bonifazi and Salvarani 1976, 137):

The agreement concluded on 16 July 1960 for workers in the
industrial sectors abolished the old pay scales differentiated
according to sex . . . and provided for a single classification.
(Commission of the EC 1979, 73)

This effectively established equality of pay for 'equal work' and 'work of equal
value'. The process required time of course, and three phases were envisioned
in the agreement. By January 1962, following a reclassification of positions, all
workers employed in equally valued positions were to receive identical wages
independently of gender. In 1962, an agreement on equal pay for nonmanual
workers was signed and, in 1963, a similar agreement covered nonindustrial
sectors (Beccalli 1984, 197).

The possibility of employers classifying as lower those jobs traditionally
held by women certainly remained. Between 1960 and 1970, for this reason,
"collective agreements that were renewed showed a steady reduction in the
number of pay categories and abolished the lowest categories, in which
traditionally female jobs had been placed" (Commission of the EC 1979, 73).
Specifically, manual workers fell into two, from eight, categories; nonmanual
workers fell into two, from six; and intermediate workers fell into one, from
three. By 1975, unions had granted women formal equality in wages in all
collective agreements (Beccalli 1985, 433, 434).

Unions moved to ensure the automatic extension of all collective agreements to nonparticipating firms and individuals (Rubery and Fenagan 1993,
16, 17, 42).[26] With over 50 percent of all women workers outside of unions,
this further demonstrated union sensitivity and commitments to equality and
to women.[27] "Every category of employment in the industrial, agricultural and
tertiary sectors" and the great majority of workers, reported the EC Commission to Brussels after an investigation of the state of equality in Italy a few
years later, became "thus covered" (Commission of the EC 1979, 53).[28]

The possibility of employers classifying as lower those jobs traditionally
held by women certainly remained. Between 1960 and 1970, for this reason,
"collective agreements that were renewed showed a steady reduction in the
number of pay categories and abolished the lowest categories, in which
traditionally female jobs had been placed" (Commission of the EC 1979, 73).
Specifically, manual workers fell into two, from eight, categories; nonmanual
workers fell into two, from six; and intermediate workers fell into one, from
three. By 1975, unions had granted women formal equality in wages in all
collective agreements (Beccalli 1985, 433, 434).

The impact of such advancements was, expectedly, phenomenal. Overall
wage differentials in industry for manual and nonmanual workers decreased
by at least 15 percent in the period 1957–63.[29] There was no major event in
the labor market in this period that could have explained such narrowing
other than the contracts. By 1970, the time from which reliable official
statistics of wage differentials in Europe are available, Italy had the lowest
wage differential of all major EC members, almost all of which were still
struggling to establish equality in agreements or law (Commission of the EC
1979, table 10).[30]

The collective agreements of the 1960s, standing at the pinnacle of a long
history of trade union openness toward working women and willingness to
represent them at the bargaining tables, made transposition of the directive
possible. They represented a labor market already respectful of wage equality:

put differently, because of collective agreements, the directive asked for a principle that would not have dramatically affected the Italian economy, women's wages or men's wages.

Strong Organization: The Feminist Ideology of the 1970s

The Equal Pay Directive (EPD) introduced ideas about equality that were already present in the collective agreements of the 1960s. This greatly helped the transposition of the directive. However, the EPD also challenged another aspect of the organization of working women. A portion of organized women in the 1970s found part of their strength and resources in a new feminist ideology that was suspicious of ideas of equality. The feminist ideology was under formation in the 1970s: the EPD's challenge concerned therefore only a limited number of women. It was sufficient, however, to ensure that trade unions would not mobilize to pressure the state to undertake the construction of a new administrative apparatus during the 1977 debates or later.

The new ideology asked women to assert their differences.[31] The hierarchical and marginalizing structures of unions came under attack early on. This happened in spite, or perhaps because, of unions' unique historic openness toward women (Beccalli 1985, 439; Cook 1984). Scorning traditional positions of leadership, women asked for the creation of independent subunits. Hesitant at first, union leaders decided eventually to honor such requests. 'Collectives', or small groups open only to women dedicated to the discussion of their problems, were thus formed.[32]

By 1977 the collectives were coordinated by a national committee. Eventually, all three major union organizations, the CGIL, CISL, and UIL,[33] had independent women committees that continued to exist into the 1990s. Feminist ideas, "the critique of the family and gender roles, the emphasis on sexuality and emotion," were by the late 1970s officially "accepted into trade union discourse," an event that "a few years earlier would have been unthinkable" by many people (Beccalli 1984, 202).

The collectives proved an important organizational basis from which to approach a new series of topics. These included female identity, self-consciousness, abortion, social services, divorce, and sexuality.[34] Legal victories in these areas encouraged the pursuit of parallel collective agreements. Laws 898 of 1970 on divorce and 877 of 1973 on the recognition of domestic work, especially, spurred a new wave of demands (Beccalli 1985, 446; Battiston and Gilardi 1992).

In fact, unions actively pursued only some of the new goals in collective agreements. They won 150 hours of employer-sponsored education courses that covered women-specific issues, such as health-care during pregnancy.[35] They also won some industry-level part-time regulations specific to women.

National, industry and local level collective agreements on maternity leaves, health rights, and information access were, on the other hand, not attempted.

The construction of new goals had one clear consequence: the attention of women unionists and unions was turned away from equality issues, including wages. "Not equality, but 'female difference' was the turf of the new movement" (Beccalli 1985, 446). The more theoretical concerns of middle-class, feminist workers became central; the concerns of the majority of women, like equality of pay, became secondary at best.

A Nonexistent Administrative Structure

The presence of collective agreements regulating the labor market essentially freed the state of any responsibility for creating the legal and administrative supports that could ensure equality of pay. An already overburdened state did not mobilize, in the 1950s, 1960s, or early to mid-1970s, to create administrative structures of any sort dedicated to the issue of equal pay, or to issues related to equality in general in the labor market. A number of governmental and interest group documents related to Law 903 recognize openly how matters related to gender wage differences were traditionally managed by trade unions and business organizations.[36]

Establishing an administrative structure concerned with equality would have been difficult to justify anyway. As discussed, Parliament proved a fairly fertile ground for protective laws, but not equality laws. All drafts prior to 1977 on equality failed to reach the status of law. It would have been unthinkable for the state to even contemplate building commissions or committees investigating or promoting equality of pay when there was no law on the matter and collective agreements had successfully regulated wage structures. Accordingly, prior to 1977, the Italian state had built no administrative structure assigned to the problems of gender equality. It would find it very difficult to change the situation after 1977.

Explaining Implementation

When the Equal Pay Directive came into force in 1975, Italy was in a position to claim that its collective agreements already provided for the principles of the law. When the European Commission expressed its disagreement, Law 903 of 1977 was drafted and passed without delays. A textual translation of the directive did not introduce any real innovation in the country: business, the state, and unions all acknowledged the guarantee of equality of pay for work of equal value in past collective agreements. The minister of labor, in a circular to regional branches, trade unions, and employers' organizations described how the "principle of equal pay has been the object of . . . collective agreements and union action," and how, therefore, the "law was to be seen" as "a means to translate, in the legal sphere of market regulations, EC directive of 9 February 1975 on equality of pay."[37] Months after the approval of the law, the Confederazione Generale dell'Industria Italiana, Italy's national association of employers, noted, in a circular to all members, how

> the principle of equality in pay and criteria for professional classification [had been already guaranteed] for years by collective agreements in all sectors of industry.[38]

In Parliament, there followed no discussions, questions or reports for the whole year after the law's enactment, suggesting that the law indeed introduced so little novelty that no one should inquire about its application, no disagreements would arise between affected groups, and no official request be made to branches of the government to account for their share of application.[39] Finally, in the opinion of critics, the law brought together an "heterogeneous body of pre-existing norms on the basis of one inspiring example" (Galoppini 1980, 262).

Administratively, the directive imposed far more extravagant demands on the state. The success of unions in representing women in wage equality collective agreements, and their failure to obtain laws in Parliament to the same effect, gradually induced the state to view unions as responsible for applying the principle of equal pay in the labor market. When the directive asked that responsibility be shifted, and that the state monitor equality, disseminate information, and sanction violators, it passed to the state a formidable and practically impossible task.[40]

Criticized for inaction after passing Law 903,[41] the government began some institutional building with the creation of a series of committees. These committees, however, had no place in the administrative tradition of the Italian system and were therefore given extremely low priority:[42] Italy, unlike France, did not have a tradition of state interventionism in the sphere of female labor regulation.

In 1982, two 'ghost committees' were instituted by the Ministry of Labor to oversee the implementation of the law. Their supposed functions included information dissemination to workers, legal support to applicants, and research activities. Specifications included personnel structures and a budget. However, no one was charged with overseeing their actual creation (Gaeta and Jannelli 1993, 136). They followed an earlier committee, set up under the Ministry of Labor by a ministerial regulation dated 17 December 1973, that had survived only one year and was brought down by an action of an excluded trade union in the Administrative Court (Commission of the EC 1979, 88).

One real committee perished within months after its birth. The Committee for the Attainment of Equality was created in 1983 and began functioning in 1984. It was made dependent for its survival on ministerial decrees: each new minister had to reinstate the committee, redefine its structure, and hire its members (Gaeta and Jannelli 1993, 136). These requirements doomed its existence and impact from the start. Given the quick turnover of ministers, the committee eventually died from 'bureaucratic overload'; it was from the beginning, a creation capable of initiative but unable to sustain any long-term project (Gaeta and Jannelli 1993, 137).

Matters would have been different had organized women pressured the state to undertake some serious administrative building. Given their feminist agenda and a change in their organizational focus, however, they remained silent during the parliamentary debates for the law. No objections were raised by anyone against the weak applicative measures detailed in the drafts.[43] Unions remained "essentially extraneous to the process of formulating" Law 903 (Borgogelli 1987, 257n5). No feminist organization within unions, stated Teresa Panariello of the National Committee for Equality in an interview in July 1994 with the author, "set to work to influence, positively, the application of the law"; fully focused elsewhere they failed "to mobilize for equality" (Beccalli 1985, 446). Outside of Parliament no formal proposal for amendments was produced by union representatives.

Accordingly, all applicative measures were designed to minimize disruptions to existing state practices. Punishment for infringements was minimal. The scope of possible repercussions was deliberately limited to single individual applicants (Beccalli 1985, 452). Unions, disinterested in the subject, accepted the responsibility of application even if equality was an outmoded topic and the law offered no concrete guidelines. All present agreed on the draft, and the law, a symbolic gesture to the EC and to all involved, was passed, unlike most national bills, unanimously in record time. Even after 1977, the EC Commission found that in Italy, contrary to Britain and most other EC members, no "parliamentary discussions, questions or reports" followed the approval of the law (Commission of the EC 1979, 33).

In sum, the implementation of the EPD in Italy can only be understood in light of institutional variables. The EPD asked for a redistribution of resources in favor of women that was already taking place thanks to existing collective agreements. Transposing the directive therefore was an almost neutral matter for state agents, business, or unions. Under EC pressures, Italy quickly and accurately put the EPD into national law.

Applying the law to close completely wage differentials, on the other hand, would have required administrative building and intervention on the part of the state in an area which had traditionally been left in the hands of trade unions and business. The state could not suddenly engage in such activities. Women, informed by a new wave of feminism and satisfied with recent past achievements, were using their organizational bases to achieve advancements in other areas and put no pressure on a disinterested state.

4

The British Case

The Weak Organization of Working Women

In contrast to the situation in France and Italy, Britain encountered enormous problems in all phases of the implementation of the Equal Pay Directive. For years after the directive's enactment, officials held that the Equal Pay Act (EPA) of 1970 adequately complied with the spirit of the EC law. In reality, the EPA had a number of serious shortcomings, including the absence of an 'equal pay for work of equal value' clause and dubious applicative measures. After losing to the European Commission in the European Court of Justice in July 1982 Britain hesitantly drafted the Equal Pay Amendment Regulations of 1983. The measure removed only some of the original EPA's flaws.

In this chapter, I argue that, due to the historical weakness of working women, Britain had built a discriminatory and stable policy legacy opposed to the progressive principles of the EPD, centered around an enormous wage differential that had for long served as a basis of Great Britain's industrial competitiveness. The directive came and asked for the impossible: the introduction of novel legal and administrative principles and the overhaul of a wage differential that would have greatly helped women but damaged the country's industrial base. The state, with the backing of the predominantly male-oriented trade unions, successfully opposed the EPD from the start.

Time and Extent of Transposition and Application

The British Equal Pay Act was enacted in 1970, but did not enter into force until December 1975.[1] The Community introduced the Equal Pay Directive in 1975. In 1975, British officials needed to adjudicate whether the EPA complied with the directive. Their opposition, during the drafting of the directive in Brussels, to 'work of equal value' clauses was an early indication of their intentions. Expectedly they declared the EPA an appropriate version of

47

the EPD in all respects (Hoskyns 1986, 309). Until 1983, following the victory of the Commission against Britain, the EPA was the official version of the directive.

The EPA failed to transpose the directive in two different ways. Its first shortfall was its narrow definition of pay. Section 6 excluded provisions made in connection with death or retirement from the concept of pay. Article 119, in contrast, specified pay to be wages or salaries and any "consideration, whether in cash or kind" received for employment.[2] Retirement schemes were included in the concept of pay, as proved in a case in 1981.[3]

The second shortfall was the absence of an 'equal value' clause. Comparisons with male employees were valid only if the jobs were identical or broadly similar. Comparisons between different jobs were allowed solely in cases where the employer had decided, *voluntarily*, to rank the two jobs using an evaluation scheme that he/she designed (Equal Opportunities Commission 1994, 20). The applicant could not ask the industrial tribunals to "order the employer or trade union to carry out job evaluation exercises" (Davies 1987, 28).[4] In effect, the rights to challenge renumeration were limited "to situations where a specific man was doing 'like work'" (Davies 1987, 30).

In light of these shortcomings, in April 1979 the European Commission initiated procedures against the British government. It argued for the violation of Article 1, since the EPA did not provide for equal pay for work of equal value. In May 1980 the Commission delivered a reasoned opinion that was dismissed by the government.[5] The Commission commenced infringement proceedings before the Court in March 1981 demanding that the concept of 'work of equal value' be part of the EPA.

The Court's decision was given on 6 July 1982. In February 1983, the government circulated for comment a draft of a new set of regulations. The Equal Opportunities Commission (EOC) of Britain criticized it, since "much of what was proposed would hinder women in claiming their right to equal pay" (Equal Opportunities Commission 1994, 25; 1993). On 6 July a minorly revised draft reached Parliament. Some in the House of Lords criticized the measure for not including the real concept of equal pay. On 16 December Lord McCarthy and other MPs attempted, without success, to persuade the government to withdraw its proposals.[6] On 1 January 1984 the Equal Pay (Amendment) Regulations entered into force.

On the positive side, the regulations expanded the definition of pay to include pensions, fringe benefits, and bonuses. Section 1(2)(c) included a clause on 'equal pay for work of equal value'. They asserted that an independent expert produce, upon request, job evaluation schemes. However, independent experts could not be summoned if the employer could show *beforehand* to an industrial tribunal that differences in pay were due to genuine differences in "material factors" between the jobs.[7] Sections 2A(2) and (3) permitted the employer to rank, unchallenged, as lower those occupations mostly performed by women. Employers could thus "persuade the tribunal that there [were] no reasonable grounds for determining that the work is of equal value, *before* . . . any expert evidence as to the value of the work is obtained" (Equal Opportunities Commission 1994, 37, italics added). The only real improvement the regulations introduced was hence an expanded definition of pay and a *conditional, narrow* definition of work of equal value.

With all their shortcomings the EPA and the amendments of 1983 still had to be applied. Prior to 1983, the government chose to rely on individual women to speak out, employers to produce voluntarily evaluation schemes, and industrial tribunals to adjudicate claims. No plans for information distribution, support, monitoring, or enforcement were made (Hepple 1987, 146–47).

The formula of relying on women and tribunals to apply the EPA proved unsuccessful. Industrial tribunals turned out to be costly. The EPA made no provision for legal aid, deterring applicants from pursuing a course of action (Szyszczak 1987, 55). Self-representation greately decreased chances of success. Tribunals proved incompetent in gender discrimination matters. Judges and panel members had little or no experience with discrimination and were mostly males.[8]

In court women faced two additional obstacles. First, they carried the burden of proof: the employee had to ask the employer to produce reasons and documents relating to her wages. Secondly, at least 25 percent more steps than the average labor case were necessary for most disputes.[9]

Most importantly, the application of the EPA was limited by section 1(6). It stated that equal pay applied only to male and female workers employed by the *same employer or from the same company* (Bergamaschi 1989, 159). Therefore, "individuals' claims could arise only within pay structures and not between them" (Davies 1987, 29). In case of victory, finally, *only the applicant*, and not all women in her condition, would benefit from the ruling.

The Amendment Regulations of 1983 sought to remove some of the applicative shortcomings of the EPA. To ensure fairness, independent experts, rather than employers, would produce job evaluations.[10] The EOC, created in 1975 as a response to the EC Court's decision against Britain for breaching the Equal Treatment Directive and placed *outside* the Ministry of Labour as an independent agency, could now provide professional advice.

Tribunals still remained the main avenue for complaints. The burden of proof remained on the woman. Section 1(6) still required that an actual worker of the other sex be employed in the same pay structure of the applicant and that victory *apply only to the specific employee.*

Independent experts increased the chances of victory. However, the average time, between 1984 and 1990, for a case to proceed from the lodging of a claim to the appointment of an expert was 19 months (Equal Opportunities Commission 1994, 51). Reports took between 3 and 17 months for completion (Hepple 1987; Equal Opportunities Commission 1994, 51).[11] Once submitted, the report could be contested again by the employer. When no appeal was made, the average decision time was 46 months. With the conclusion or withdrawal of appeals, that time grew to 60 months (Equal Opportunities Commission 1994, 52). The EOC and both houses had criticized the regulations before their enactment:

> No ordinary lawyer would be able to understand [these regulations]. The industrial tribunals would have the greatest difficulty and the Court of Appeal would probably be divided in opinion. (Lord Denning, Hansard (HL) 1983, cols. 901–2)

> Not only is it wasting the time of the House to induce a measure of this kind, but . . . it will almost certainly result in further action being taken in the European Court, which will again prove this country to be in breach of its obligations to women under the equal pay directive. (Mr. Robert Maclennan, Hansard (HC) 1983, cols. 496–97)

> [This] is legal gobbledegook. It is an algebraic mystery . . . [an] obtuse maze of a measure and truly a stumbling block to a female compliant and her advisers. (Mr. Barry Jones, Hansard (HC) 1983, col. 497)

Why, then, did Britain resist the implementation of the EPD in 1975? Why were the 1983 regulations so inappropriate?

Unions and the Depression of Wages

The EPD could not be implemented because, as a result of women's underrepresentation in the labor market throughout the twentieth century and the consequent depression of women's wages, Britain built a discriminatory policy legacy reliant on cheap female labor opposed to the progressive principles of equality. The state, backed by conservative unions, refused to suddenly undermine its entire economy. This section analyzes the nature of women's underrepresentation and the consequent depression of female wages.

From their birth, British unions acted as poor representatives of female workers.[12] Between the early 1880s and 1950, unions openly *endorsed* wage differentials. Between 1951 and 1970, unions became partly supportive of only *equal pay for equal work*. Table 4.1 outlines underrepresentation in the British labor market. It identifies levels of female membership and leadership in trade unions, and the major policies pursued by organized women during the period 1820 and 1970. The following pages elaborate on all three columns.

Until 1888, as descendants of seventeenth- and eighteenth-centuries male guilds, unions simply refused to accept women in their ranks or to fight on their behalf (Cook 1984, 12). Official policies of exclusion, such as the one presented at the National Conference of the Cotton Spinners in 1829, were in fact signed throughout the kingdom (Soldon 1985). By 1880 women were excluded from all trade unions but one.[13] They were encouraged, instead, to form separate, all-female organizations (Lorwin and Boston 1984, 141). Justifications for such behavior are well documented; they included discriminatory conceptions of the role of women, arguments about inability to endure strikes, and explicit fears of wage competition (Soldon 1985, 15).[14]

Toward 1900, female presence in some industries reached 50 percent of the workforce. Recognizing that "active opposition to women's membership in unions became neither practical nor desirable," unions created second-class, nonbeneficial, low-dues forms of membership, a move, according to preeminent union historians, that

> sanctioned separate low-wages scales for women and discouraged women's participation in these organizations by seriously limiting their rights and scope within the unions. (Cook, Lorwin, and Daniels 1992, 22–23)

Only one union prior to 1910 admitted women in numbers proportional to their presence in the labor force, the Dundee Jute and Flax Workers Union in 1906 (Lorwin and Boston 1984, 151). Limited participation meant no or merely symbolic presence in the national executive committee. Until 1920, when the Women's Trade Union League and other such leagues joined the TUC with the understanding that women hold two seats in the General Council, there were no women in high positions of power.[15]

Viewing requests for increases in wages as a threat to male wages, unions opposed such demands and sought the segregation of women into low paid jobs. The speech of a male unionist in the printing industry exemplified this widespread sentiment:

> [Since] women are not physically capable of performing the duties of a compositor, this Conference recommends their admission to membership of the various typographical unions upon the same conditions as journeymen, provided always the females are paid strictly in accordance with the scale. (quoted in Soldon 1978, 46)

In 1888, the TUC passed a resolution on equal pay for equal work.[16] The motion was adopted, however, explicitly "to dissuade employers from hiring women," who at the time were flooding the market (Soldon 1985, 14;

Table 4.1. Women's Underrepresentation in the British Labor Market, 1820–1975

	Female Union Membership	*Union Leadership Positions*	*Major Policies Pursued by Unions*
1820–1880	Complete exclusion except textile union	None	None
1880–1914	Nonbeneficial membership	Almost none; TUC delegates	Protective legislations
1914–1918	Increases, as men leave for WWI	Almost none	Agree with War Cabinet Committee to protect men's jobs/wages
1918–1939	Back to pre-war levels	2 women in TUC General Council	Full employment for males
1939–1950	Increases again	"	Protection of male wages
1950–1961	Fairly high as economy booms	"	Equal pay for equal work an issue, esp. in public sector
1961–1975	Steady	Enlarged but still one of the lowest in Europe	Along with Labour Party, TUC withdraws support of equality of pay

Bergamaschi 1989, 134). The TUC's reasoning, which proved correct, was that employers would choose men's labor over women's for similar costs.[17] The resolution was not, in any case, followed by any national-level or industry-level agreement, or legal initiative. It left to individual unions the responsibility of promoting equal pay.

Independent organizations for women were born as a reaction to the conservatism of unions. They included the Women's Protective and Provident League,[18] the Women's Trade Unions League (WTUL), and the impressive National Union of Women Workers (NUWW).[19] They pushed the TUC to fight for harmless 'protective' legislations and not equality: laws prohibiting night-work and overtime, lifting weights, standing long hours, and working underground (Cook, Lorwin, and Daniels 1992, 22–23), and the Trade Boards of 1909, for minimum wage floors for all workers (Bergamaschi 1989, 135).

The ideological bases of the formative periods, so different from the Italian case, shaped the response of unions to the influx of women into the labor market during the two world wars. The influx was similar for both wars. World War I brought 1,345,000 women, mostly in textiles and munitions; union membership grew from 357,956 to over 1,000,000 between 1914 and 1918 (Soldon 1978, 80). World War II brought 2 million women into factories for a total of 9 million. Female union membership again increased to 1,219,543 (Soldon 1985, 23). In both cases, at least one third of all new working women replaced men, and in both instances unions acted to protect men. Most of their efforts went to ensure that men would replace women upon their return from war, and that men's wages would be the same as before the war.[20] Wages in their view could either drop (as long as women held the job) or remain altogether unaltered. Thus, unions signed with the War Cabinet Committee the Pre-War Practices Acts of 1919 and 1942: these assured men that female jobs acquired during the war could be vindicated after the war and that women's "salaries should be inferior to those of men" in all industries, with the apparent exception of the munition industry (Bergamaschi 1989, 136, 154–55).[21] An Equal Campaign Committee organized by women in the TUC in 1944 received little attention and promoted no policies (Smith 1981).

Unions behaved in the same way between the wars. General high rates of unemployment, following economic depression, and not women, received the TUC's attention. Gender issues lost relevance even in the traditional "bastions of women trade unionism," as the textile unions, where female membership strongly declined (Soldon 1985, 20).[22] Numerous alternative all-women unions and groups merged with the TUC and obtained the right to annual conferences (the Women's Conference), but they could only produce advisory motions and no policy proposals. The merge served primarily to relieve the TUC of an increasing worry with female unrest and defection.[23] The few, occasional equal pay motions went unheeded, as in 1938 at the Congress of the National Union of General Municipal Workers.[24]

Soon after the Second World War, the TUC, at its annual conference, strongly reasserted in words and deeds its position concerning women:

> Home is the most important sphere for working women . . . and it would have been a grave mistake for the life of the nation if women were convinced or forced to neglect their domestic duties to enter industries. (TUC 1947, 247)[25]

The TUC rejected an attempt to change the Women's Conference from an advisory to a policy-making body.[26] It also rejected the attempt to approach the government with a proposal for establishing equal pay, reasoning that "a further approach to the government would be inappropriate at the present time."[27] As time went by, however, the great increase in women's participation in the labor market and union membership (16% in 1945, 29% in 1970) began at last to shake the conservatism of the unions.[28]

Ideas about 'equal pay for equal work' started to circulate throughout much of the TUC. On two important occasions in 1963 and 1965, the TUC's General Council responded somewhat favorably to proposals by the Women's Committee, such as the *Industrial Charter for Women*, to pursue concrete action for equal pay in the industrial sector (TUC 1965, 415). The rhetoric within the Congress heated up. Women, a frustrated union leader said, earn "little over half of what is paid to men for similar employment"; it is time, he added, that "something was done about it" (TUC 1965, 414). That year, a plea was made to the president of the TUC: "What an extra bonus to the [TUC's] centenary celebration could it be if . . . we could say that . . . we had made substantial and proper steps toward the achievement of *equal pay for equal work*" (TUC 1965, 415, emphasis added). Almost no one dared to introduce in the discourse 'equal pay for work of equal value'.[29]

Public sector unions accepted the pleas. Their unions mobilized, successfully, for equality of pay for identical jobs in 1961 (Lawrence 1994; Meehan 1985, 43). The rest of the TUC resisted them. When the Labour government postponed considerations of equal pay to more favorable economic times, Minister of Labour Ray Gunter was invited to explain to various unions how equal pay raised such complex issues that industry could not be expected to introduce it. In a momentous decision, the TUC eventually *dropped* its support of national legislations for equal pay, favoring local bargaining instead, and restated its loyalty to the philosophy of the party.[30] The decision to follow the lead of the Labour Party spurred "angry scenes at the TUC conference on equal pay when TUC leaders were accused of having been hoodwinked for too long by the government" (Meehan 1985, 39). Frank Cousin, the general secretary of the Transport and Workers' Union expressed frustration at the TUC's turnaround:

> We have given a tremendous amount of lip service over the years to the problem of the lower paid . . . if men wanted equal pay for women, they would get it. (TUC 1967, 534)

He and others insisted that the TUC should press the government to implement its promises (Soldon 1978, 176). The TUC responded by twice sending a team made up entirely of men to speak to the government and to business. This happened even though two women were on the General Council and the Women's Advisory Committee had complained to the General Council after the first had been sent.[31]

The decisive moment came at the 1969 yearly conference. First, Prime Minister Harold Wilson was given a chance to remind leaders that equal pay would mean fewer increases for men (Bergamaschi 1989, 142). Then, "a number of unions represented" spoke up and "overtly opposed equal pay" for *identical* work (Cook, Lorwin, and Daniels 1992, 98; Soldon 1985, 26). They were, in the words of a union member, still unsure about "the appropriate definition" of terms, "presumably according to whether their main aim was to safeguard the position of their male members or to gain parity for women" (TUC 1970, 7).

Encouraged by protests and a strike at the Dagenham plant of the Ford Motor Company,[32] a frustrated new Minister of Labour, Barbara Castle,[33] leaders of women's organizations, and radical trade unionists[34] agreed to move the cause "beyond unions,"[35] openly denouncing unions' "discriminatory attitudes."[36] Their efforts eventually led to the 1970 EPA.

By then, however, the damage had been done. The absence of representation in collective bargaining quite naturally led to the depression of women's wages over the years. By 1955, women overall were earning only 53 percent of men's wages, possibly the lowest percentage in Europe. Table 4.2 shows how, between 1935 and 1960, only minor advances were made in wages in those sectors with traditional heavy female participation, namely the lower professionals (55% presence), clerks (60%), and semiskilled manual workers (41%). By 1972, for all industries and services, women manual workers were earning 62% of male manual workers' wages; women nonmanual workers were earning 53% of male nonmanual workers (Commission of the EC 1979, table 10). Such unyielding high wage differentials had become extremely valuable to British industry and, by reflection, to the state.

High Costs: The Historical Opposition of Government to Equal Pay

The Equal Pay Directive entered one of the most antagonistic policy-legacies in Europe. Both legally and administratively, the British state had little to boast in 1975 in terms of support of gender equality in the labor market. Labour and Conservative governments alike, concerned with the consequences of equal pay on the well-being of British industry, had opposed any progressive law throughout the postwar period up to 1975.[37] Government reasoning was straightforward: given the wide wage differential between men and women (that had resulted from years of underrepresentation), and given Britain's low labor costs and poor international competitiveness, closing the differential would have greatly harmed British industries as well as the national economy. Accordingly, free from any union pressure to do otherwise, the state would introduce no legal or administrative entity that would support equality of wages in the labor market.

The gap between men's and women's wages was indeed large throughout the postwar period and up to 1975. Years of underrepresentation had ensured the depression of women's wages.[38] Table 4.3 shows the wage differential in comparison to that of other major EC countries between 1969 and 1975, the years for which reliable data exist. The wage differential became increasingly

Table 4.2. Women's Wages as Percentage of Men's Wages (%w) and Women's Presence (%p) in the Labor Market over Time in Great Britain, 1935–1960

Year	Unskilled Manual %w	%p	Semi-Skilled Manual %w	%p	Skilled Manual %w	%p	Clerks Administrative %w	%p	Managers Professional %w	%p	Lower Professional %w	%p	Higher Professional %w	%p
1935	57		75		44		46		38		69		—	
1955	52	20	57	41	51	15	57	60	54	18	72	55	75	8.5
1960	53	22	53	42	50	16	61	62	54	20	72	61	75	8.1

Source: Data from George Routh, *Occupation and Pay in Great Britain 1906–1960* (Cambridge: Cambridge University Press, 1965).

Table 4.3. Gross Hourly Earnings of Female Workers as a Percentage of Earnings of Male Workers in All Industries in Four Countries, 1969–1977*

	1969	1970	1972	1973	1975	1977
Britain	n.a.	n.a.	59.5	61.1	67.3	71.3
Italy	73.9	72.7	77.0	77.3	81.2	84.3
France	78.4	78.3	78.7	80.4	78.3	77.4
W. Germany	69.4	69.4	70.0	70.5	72.7	73.0

Sources: Data from Eurostat, *Economic and Social Position of Women in the Community* (Luxembourg: Statistical Office of the European Communities, 1980), table 5.9; Eurostat, *Earnings: Industry and Services* (Luxembourg: Statistical Office of the European Communities, 1992), quoted in Rubery and Fagan (1993, 157).
* Manual and nonmanual workers.

important to Britain and its politicians for two related reasons. First, Britain had, partly because of the differential, one of the lowest average labor costs in Europe (see table 4.4). Secondly, postwar Britain was in an industrial crisis that magnified the value of low labor costs. Between 1950 and 1960 Britain had a steady 3% average annual increase in output, which was only 50% of the collective rate of increase of the six members of the EC; its prices were higher than everywhere else except France; and imports were increasing more quickly than exports, yielding a worrisome trade deficit. In the 1960s, industrial output grew by only 3.5% compared to a growth of 5.0% by other members of the EC. Inflation also continued to grow until it reached, in 1974, 15.5%, compared to 5% in West Germany and 11% in France (George 1985, 67).

"Keeping down hourly wage costs" became therefore a government priority (Hoskyns 1996, 313).[39] All labor costs were targeted with a ceiling on wages: in the 1970s, Britain was the only European country to show no real wage increase over time (Eurostat Review 1980, 148). The equalization of wages became, however, especially feared: *closing such a large gap would surely cause wages to shoot up in a time of industrial crisis.* Estimates of the impact ranged between 3% and 15% of production costs.

Thus, before 1970, Labour and Conservative leaders alike worked to oppose and then limit any effort toward equality (Davies 1987, 30). In 1946, the government published a report describing the inflationary effect of equal pay.[40] In the early 1950s, the government rejected all proposals for equal pay originating from the TUC's Women's Conference: Chancellor of the Exchequer Hugh Gaitskell, for example, opposed the equal pay measures proposed by the National Union of Bank Employees "for the common classes" of the public service (TUC 1950, 454–58; Soldon 1978, 162). Only in 1954, following meetings and the gradual extension of equal pay in many local

Table 4.4. Hourly Labor Costs in ECU in Six Countries, 1970–1978*

	1970	1971	1972	1973	1974	1975	1976	1977	1978
Britain	—	—	—	2.11	2.51	2.95	3.19	3.49	3.81
Italy	2.19	2.28	2.57	2.89	3.39	4.20	4.59	5.35	—
France	2.15	2.16	2.60	3.15	3.58	4.59	5.36	5.87	6.51
West Germany	2.68	3.00	3.44	4.28	5.13	5.76	6.74	7.77	8.56
Netherlands	2.51	2.81	3.40	4.15	5.36	6.45	7.58	8.34	9.03
Denmark	—	—	—	—	—	5.74	6.68	7.15	7.69

Source: Data from Eurostat, *Review 1970–1979* (Luxembourg: Statistical Office of the European Communities, 1980).
* Manual and nonmanual workers.

city governments, did the chancellor, this time R. A. Butler, decide to accept equal pay in the civil service (Soldon 1978, 172).

The toughest ground was the private sector. Between 1960 and 1965 the strongest pressure came from the TUC, led by the Women's Conference, which asked, however feebly, government to comply with its pledges and with ILO Convention 100,[41] and warned that equal pay for the same work would be required under Article 119 of the Treaty of Rome should Britain join the EC.[42] The Ministry of Labour refused to consider the matter, and replied: "equal pay was not a matter for government intervention but for industrial negotiation."[43] He explicitly referred to the 1946 *Report of the Royal Commission on Equal Pay* and its findings that equal pay would be inflationary (Soldon 1978, 161).

Once in place in 1964, the Labour government "faced the same economic problems as its predecessor" (Soldon 1978, 175). Its new National Board for Prices and Incomes decided to limit wage increases to 3.5 percent, with the exception of areas in production where workers were extremely ill-paid. The new minister of labour, Ray Gunter, opted not to include women among the exceptions; instead, he "appointed an inter-departmental committee to consider its effects" (Soldon 1978, 175). In May 1965, Gunter hesitated to take further action: he kept the findings secret, fearing that costs would have been too high for Britain.[44]

Gunter's parliamentary secretary confirmed to concerned unions and women's pressure groups later that all election pledges would be fulfilled. The minister of labour himself pledged at the Labour's Women's Conference that the government would seek ways and means of implementation (TUC 1965, 413). He nevertheless was instrumental at setting up a tripartite study group, composed of unions, management, and government, in 1966 to discuss costs. By the middle of 1967, talks were deadlocked over definitions and estimates of costs.[45]

Barbara Castle replaced Gunter at the end of 1967 and immediately commissioned new studies to consider costs in specific industries and departments. Although she was clearly in favor of wage equality, other important figures in the government opposed it. In 1969, Prime Minister Wilson warned trade unions that higher wages for women would imply lower wages for males, pointing out that the cost of implementing the EPA would range from 2% to 13% (32% in a few cases) of total current production costs, and that such increases would freeze men's wages. Many in the ministry of labour openly feared that an investigation of wage-setting criteria would reveal inconsistencies among male wages and would, in the words of a civil servant involved in the design of the EPA as a representative of the ministry, "upset the pay differential between male workers."[46] Again, government opposed any wage increase: a new act in 1968 had imposed a ceiling of 3.5 percent on all wage increases and empowered the Price and Incomes Board to postpone all increases for twelve months (Soldon 1978, 176).

Facing an antagonistic environment, Castle had to introduce great limitations to the EPA. All proposals to enhance the role of industrial tribunals, force employers to produce evaluations upon request, or extend the applicability of a woman's victory were rejected. Protests during parliamentary debates were of no use.[47]

Between 1970 and 1975, pressures from women's groups[48] and the ILO to conform to the upcoming Community law mounted. The government started deliberate, and successful, efforts to prevent implementation. Leaving the EPA intact, officials proclaimed it a proper implementation of the directive.[49]

Explaining Implementation

The Equal Pay Directive imposed great demands on Great Britain. In terms of the organization of women, the directive asked for a major restructuring of the labor market. Women would be awarded very high increases in wages, men would probably see their wages freeze or shrink, industrial competitiveness would be undermined. Legally and administratively, the directive asked the state to produce novel principles that had been opposed explicitly by a succession of governments throughout the postwar period.

Consistent with its past approach to equal pay and unopposed by unions, the British state thus simply insisted in 1975 that its EPA of 1970 conformed to the requirements of the directive. When, in 1982, the government was found in breach of the directive by the European Court of Justice, the same approach continued. The amendments to the EPA were introduced to the House of Commons by the under-secretary of state for employment, Mr. Alan Clark MP, who was criticized by other MPs for his obvious hostility to the principle of equal pay.[50] The Department of Employment put the proposed changes, including the unacceptable 'material factor' clause, entirely in the context of the European Court's ruling (Davies 1987, 34). *The amendments were submitted not as primary legislation but as subordinate legislation.* The use of this procedure made it impossible for members of Parliament to table amendments to the regulations; they could either vote to approve or reject the regulations as a whole.[51]

On 20 July 1983, some in the House of Commons attacked the revisions by saying that "the procedures' . . . sole object appears to be to deter the maximum number of applicants from seeking remedy and to provide the greatest possible resistance to those who persist" (Mr. Robert Maclennan, Hansard (HC) 1983, cols. 496–97). Ms. Jo Richardson reminded all present that the majority of opinions sent to the Department in the past had been critical of the government's proposals.[52] For the most part, however, these comments were isolated[53] and the government obtained the Commons' approval with a comfortable majority.[54] On 6 September, the government circulated the draft again before submittal to the House of Lords. The EOC, following the advice of the Leading Counsel, disagreed that an employer could defend his 'genuine material factor' evaluation prior to the work of an independent expert, and it criticized the judicial process as complex, obscure and capable of draining the resources of applicants, employers, tribunals, and the public generally.

The government proceeded to submit an unaltered draft of the regulations to the House of Lords anyway. The debate took place on 5 December 1983.[55] The Earl of Gowrie spoke on behalf of the government. In commending the regulations, he acknowledged the criticisms that had been made of the government's proposal, but argued that their complexity was unavoidable, and claimed, contrary to all evidence, that "the operation of the new provision can be described as fairly simple." When acknowledging some of the limitations, he offered no alternative solution. He knew that limiting victory to the plaintiff avoided "collective repercussions," and simply noted: "we have not however provided any specific mechanisms to deal with those."[56] Members denounced the government's work, deriding the revisions' "tortuosity and complexity . . . beyond compare" and describing it as a naive attempt to please the EC.[54]

In spite of all this, the House of Lords approved the regulations. On 16 December Lord McCarthy and other MPs attempted, without success, to persuade the government to withdraw its proposals.[58] On 4 February 1984, Baroness Seear, in the House of Lords,[59] and Mr. Maclennan MP, in the House of Commons,[60] attempted in vain to have the proposals annulled.

No criticism after 1983 was fruitful. In its 1989 consultative document,[61] the EOC criticized most of the shortcomings of the EPA. In its *Formal*

Proposals of 1989 and *Equal Pay for Men and Women: Strengthening the Acts* of 1990 the EOC repeated these suggestions. Between 1988 and 1993, members of Parliament asked successive secretaries of state whether they will give effect to the EOC's legislative recommendations. The replies given showed that the government had not given serious consideration to the proposals.[62] For instance, on 20 November 1992, already two years after the EOC submitted a formal proposal that included all of the above suggestions, Mr. McLoughlin MP of the House of Commons casually observed that "the Equal Opportunities Commission's recommendations for changes to the sex discrimination and equal pay legislation are receiving careful attention by the Department" but that no measures had been taken to implement any of the proposals (col. 396).[63] Nothing had changed from a year before, when the government had observed that the EOC's proposals of 1988–90 were "under consideration."[64] Only in 1993, under threat by the EOC that it would report the government's behavior to the EC Commission, was there a response. The government had rejected most proposals. By then, it became obvious for reason that we now understand that the government's "objective in the whole process [had been] to do the minimum necessary to comply with the law" (Davies 1987, 34).

In its work between 1975 and 1983, and after the 1983 regulations, the government benefitted from the silence of unions, consistently weak representatives of women's interests. Unions, under the influence of feminist currents and higher female membership,[65] were certainly beginning to experience a transformation toward a more serious commitment to equality (Lawrence 1994, 58). Internal committees and working parties focusing on equal pay, for instance, were formed. In 1977 the TUC held a special conference on equal pay. In 1977 the TUC's Women's Conference called for substantial amendments to the Equal Pay Act. In 1979, the TUC adopted the Charter for Equality for Women within Trade Unions. One of its objectives was to have women "adequately represented on union executives, that advisory committees on women's issues [be] established, that meetings [be] held at convenient times for women" (Lawrence 1994, 33). By 1981, reports from Women's Advisory Committees were flooding the TUC headquarters urging for more severe action on the part of the TUC. The TUC, after much hesitation, even sent the secretary of state a copy of proposed amendments to the EPA in 1978 that included the notion of 'work of equal value' (TUC 1978, 86).

For the most part, however, the principle of 'equal pay for work of equal value' was hardly a real union goal. Most of the progressive work was still aimed at convincing union members and leaders that women had a right to equal pay for equal work. Women from fifty-eight unions, for instance, organized the new Coalition of Labor Union Women in the late 1970s to persuade "cautious unions into unequivocal opposition to discrimination" (Meehan 1985, 95). A number of studies found widespread opposition and hostility to the EPA among unionists in general. In 1973 and 1974, one such study found, numerous unions officials were even opposed to women working in general (Meehan 1985, 41). A second study found that in 25 percent of private companies the application of the EPA had actually "been blocked by the attitudes of the male union members," who were in some cases caught manipulating pay scales (Meehan 1985, 96).[66] For most union representatives and officials, a third report found, "equal pay was a minor issue, peripheral to their central concern and worries."[67]

The TUC never truly mobilized a large-scale initiative, in the form of documents, policies or strikes, to ensure the proper transposition or application of the EPD between the crucial years of 1975 and 1983. It never forced the government to take up the question of whether it had properly implemented the EPD until the European Court mobilized. In 1983, the government deliberated over the amendments of 1983 freely, without pressures from unions. Such silence from unions was hardly surprising: it only confirmed a hundred years of prior antagonism to the cause of working women.

In sum, the EPD would have introduced changes beneficial to women. But women could never ensure that a progressive EPD would replace a limited EPA, itself the product of a government firmly opposed to the principle of equal pay. The government, requested by the EC to alter its policy on women's work, was at liberty to bypass opposition, ignore the EC, and defend what it perceived to be the interests of the nation.

The British case shows perhaps most clearly the importance, for implementation, of the directive's demands on the organization of interest groups and national policy legacies. In part II of this volume, the same institutional approach will be used to explain the fate of a second directive in Spain, Italy, and Great Britain.

Part II

The Sulphur Dioxide and Suspended Particulates Directive

Prologue: The Directive

The 1980 Sulphur Dioxide and Suspended Particulates Directive (SSD) was the first Community measure taken to harmonize air pollution regulation in Europe in order to protect the health of its citizens and to ensure that industries in different countries would be subject to identical penalties and constraints.[1] The Commission presented it to the Council in 1976. It tackled one of the most serious forms of air pollution: sulphur dioxide and smoke (suspended particulates) originating mostly from industrial and power stations' emissions.

Prior to 1972, the European Community did not have an environmental policy. The 1957 Treaty of Rome, which established the Economic Community, did not mention the environment as a policy issue, probably because at the time problems linked to environmental abuse, such as acid rain and transboundary pollution, had not occurred.[2] Since 1973, following the 1972 Paris Summit of the Heads of Governments, four-year-long action programs have specified increasingly concrete objectives and strategies for the protection of the environment. Major events, such as the Single European Act of 1986 and the Maastricht Treaty of 1992, have made environmental protection a major objective, consistent with economic growth. The Treaty of Rome was modified to include articles on the environment and structures, such as a directorate-general within the Commission and the European Environmental Agency, were set up to design and pursue policies.[3] As a result, over two hundred directives today protect and regulate the pollution of air, water, flora, and fauna.[4]

Although proposed in 1976, the directive was adopted by the Council on 15 July 1980, after extensive debates and negotiations.[5] The Commission

Table II.1. Limit Values of SO₂ and Suspended Particulates as Specified by Directive 80/779

Period	Limit Value for SO₂ (mg/cubic meter)	Limit Value for Smoke (mg/cubic meter)
Year (median of daily values)	120 if smoke ≤ 40 80 if smoke > 40	80
Winter (median of daily values, Oct.1- Mar. 31)	180 if smoke ≤ 60 130 if smoke > 60	130
Peak (98th percentile of all mean daily values throughout the year)*	350 if smoke ≤ 150 250 if smoke > 150	250

*These values, moreover, cannot be exceeded for three consecutive days.

asked that the directive be transposed by 17 July 1982 and its principles be applied by 1 April 1983.

The directive's main purpose was to limit the presence of sulphur dioxide and black smoke in the atmosphere in order to safeguard human health and the environment, and to ensure that countries already active in this field, such as Germany, would not be alone in spending resources and penalizing domestic industries.[6] To achieve these goals, the directive stated five principles, all of which had to appear in national legislation and be applied.

First, the directive imposed 'limit values' on the concentrations of sulphur dioxide (SO₂) and smoke in the air for the whole year, for winter periods (1 October to 31 March), and for peak periods. The limits of SO₂ and smoke, expressed in micrograms per cubic meter of air, varied in relation to each other, as laid out in table II.1. National laws had to include these values and states had to ensure that they be respected by 1 April 1983. Secondly, the directive enunciated the 'standstill principle'. It stated that air quality could not deteriorate significantly even in areas well below the limit values. Again, national laws would need to have a provision stating this principle, and states would need to take measures which ensured that identified areas would be protected.

Thirdly, the directive asked that stations be established throughout the territories, and especially in those likely to violate the limit values, to measure pollution levels. National laws had to state that such stations were in fact needed, and states had to provide for the establishment of those stations.

Table II.2. Implementation Records of Spain, Great Britain, and Italy for Directive 80/779

80/779 Required:	Limit values	Standstill principle	Stations for sampling	Guide values	Compliance plans for exceptions
Spain	Transposed on time, partly applied	Transposed on time, partly applied	Transposed on time, partly applied	Transposed on time, partly applied	Transposed on time, partly applied
Great Britain	Transp. very late, mostly applied	Transp. very late, mostly applied but late	Transp. on time, applied	Transp. very late, mostly applied but late	Transp. very late, mostly applied but late
Italy	Transp. very late and in part, never applied	Transp. very late, never applied	Transp. very late, never applied	Transp. very late, never applied	Transp. very late, never applied

Fourthly, in annex II, the directive set 'guide values', or more stringent values that would be imposed in the future. Member states were asked to specify these values in their laws and to begin planning for compliance.

Finally, the directive stated that a country not able to respect limit values (i.e., the first requirement) for SO_2 and smoke would need to present concrete measures for improvement to the Commission by 1 October 1982. The deadline for applying the directive would then be moved to 1 April 1993. This final requirement, too, had to appear in the legal translation of the directive. States for whom it applied were expected to take measures to ensure that it be respected.

Table II.2 summarizes how the requirements of directive 80/779 were implemented in each of the selected countries. Spain faithfully transposed all of the directive's requirements on time, but could only apply them in part. Great Britain transposed the directive fully only in 1990; it, like Spain, applied only some of its principles on time. Italy's was perhaps the worst record: the government transposed the directive almost fully but late (like Great Britain), but also failed to apply all of its principles. The ensuing chapters investigate closely the nature and causes of implementation in each of these three countries.

5

The Italian Case

An Unfavorable Policy Legacy

Despite a serious environmental crisis that began in the 1950s, Italy transposed the directive partly and with great delays. It also never ensured its application. The first legal measure used for implementation, the Decreto del Presidente del Consiglio dei Ministri of March 1983, omitted a number of important requirements for the immediate and future protection of the air. It also failed to specify serious measures for application. The second legal measure used for implementation, the Decreto del Presidente della Repubblica of May 1988, included four of the directive's five requirements. None of the principles, though, were properly applied.

In this chapter, I explain Italy's performance in terms of the demands that the directive imposed on the policy legacy of the state. Italy lacked, in 1980, a serious law regarding the regulation of air pollution; it had no administrative apparatus for the protection of the environment. The directive could not be implemented because it imposed extraordinary and impossible demands on the state. Once the state began to develop a serious environmental policy in the late 1980s, the legal context for the directive's transposition came into existence. Administratively, Italy was then only beginning to develop structures for the environment and could not mobilize to ensure the directive's application.

The directive's demands on the organization of interest groups engendered dynamics that were only secondary for implementation. Conservation groups were too weak to pressure the state to adopt or apply a proper law that would have, possibly, strengthened their legitimacy and resources. Producers of pollutants would have probably mobilized against the directive had there been a need. As such, the state's reaction to the directive essentially preempted producers' reactions.

Time and Extent of Transposition and Application

Italy transposed the directive partially and with great delays. It produced two decrees to claim that it had officially complied with the requirements. These were the Decreto del Presidente del Consiglio dei Ministri (DPCM) of 28 March 1983 and the Decreto del Presidente della Repubblica (DPR) of 24 May 1988, No. 203.

The first decree was an administrative, not a legal act. This could have undermined its validity as a legal translation of the directive from the start. The Commission, however, did not raise significant objections concerning this matter. The 1983 DPCM set limit values for SO_2 consistent with the directive. It also asked that, for those areas in excess of the limits, regions draw up improvement plans that would ensure compliance by July 1993.[1]

The DPCM, however, failed to transpose many of the directive's other requirements. Although it specified limit values for SO_2, it did not specify the correct limits for smoke, the winter median of daily values, or the peak values. It also failed to relate SO_2 and smoke values to each other.

The DPCM did not include the standstill principle. By failing to control presently clean areas, Italy increased the chances of future breaches of the limit values. The DPCM did not specify the need to have monitoring stations throughout most of the territory, and especially in polluted areas. It instead delegated, in vague terms, supposedly related administrative duties to different ministries and agencies. The DPCM did not mention the guide values that the directive listed in Annex II as ideal limits for the future. No measures were thus specified to begin preparations for the next phase, making it likely that Italy would breach EC law on air pollution again. With some exceptions (mention of some limit values and request for compliance plan) the DPCM thus clearly failed to transpose the directive in question.

The poor transposition of the SSD was not an isolated event in the area of environmental policy. Other environmental directives of this period experienced similar fates.[2] Facing the threat of infringement accusations from the European Commission, the Italian Parliament passed Law 183 of 1987, which empowered the government to adopt directives with decrees that carried legal validity.[3] This effectively permitted government to act as parliament: to transpose freely and decide on the means for application. It is in this light that the legal validity of the second decree, DPR No. 203 of 1988, must be understood.[4]

The DPR corrected many of the shortcomings of the DPCM (Capria 1992, 143). It introduced winter mean limits, peak limits, guide values, and the standstill principle consistent with the directive. While the government retained its control over defining the limit and guide values, it opted to delegate to individual regions the duty of monitoring, with stations, air levels.[5] With the important exception of failing to relate SO_2 values to smoke values and to adjust the smoke value limits, the DPR correctly transposed the directive.

Both the DPCM and the DPR had to be applied. Simple[6] limit values for SO_2 and smoke, along with a request for compliance plans, were the only requirements put forth in the DPCM. Because of vague and confusing administrative delegation, along with the lack of serious penalties, the decree was never applied (Capria 1992, 161–62). Accordingly, the directive itself was not applied.

The DPR, by contrast, included specific plans for application, which were in turned further clarified by a follow-up administrative decree in 1991.[7] Regions were expected to set up monitoring stations and collect data. At the national level the decree asked the minister of environment to: (1) issue guidelines for regions to design improvement plans for limit and guide values; (2) decide on the distribution of financial support; (3) identify which areas should respect the standstill principle, and which areas would need until 1993 to comply; (4) specify criteria for future data collection and reports. The decree originally granted the ministry 180 days (i.e. until 15 December 1988) to perform this duty (Capria 1992, 145, 155).

Evidence indicates that, despite such clear planning, application of the decree never occurred. Few regions built stations. Only six regions prepared improvement plans within the prescribed time line. Other regions "passed legislation to draft plans but [had] yet to decide on their contents" (Capria 1991, 69). National coordination, i.e. steps (2) through (4), and the regional activity that should have then followed were stalled from the start.

Such failure to mobilize for the protection of air quality and the health of citizens could have been overlooked had SO_2 and smoke values been under the limits prescribed by the directive. However, there is evidence to indicate that pollution levels were indeed higher than those limits. In 1986, the national energy and environment agency, Ente per le Nuove Tecnologie, l'Energia e l'Ambiente (ENEA), conducted research to identify the major sources of pollution in Italy.[8] It gathered data from a disparate number of agencies, laboratories and some monitoring networks.[9] "Despite the weakness of the material, useful conclusions could be drawn" about the concentration levels of SO_2 and smoke:

> The sample sulphur areas showed a tendency to exceed the limits values for sulfur dioxide and suspended particulates in all the large urban centers, though the concentration progressively decreased from 1978 to 1985, mainly due to controls on domestic heating fuels. (Capria 1991, 70)

Thus, Italy transposed the directive almost fully, but very late and failed to apply it, even though its requirements were being violated.

This outcome, in an area as important as the quality of air in a country, must be explained. What caused the initial, almost complete, disregard of this directive? What can explain its later partial transposition but the persistent failure to apply it? The next sections offer an answer to these important questions by focusing on the policy legacy of the Italian state on the environment.

Social and Political Responses to the Environmental Crisis

Italy experienced a serious environmental crisis in the 1950s and 1960s, following a remarkable postwar growth in polluting heavy and chemical industries, unregulated deforestation, land use, and urban development. Numerous observers at the time described the country as an environmental disaster.[10] The Organisation for Economic Cooperation and Development (OECD) showed Italy as having one of the heaviest concentrations of air and water pollutants in Europe.[11] The response to the crisis prior to 1980 was weak and came in two forms: socially, an intellectual middle-class movement and a few conservation organizations framed the problem primarily in terms of the hazard posed to Italy's cultural heritage and cities and exercised some, but generally ineffective, pressure for action; politically, parties in government and in the opposition belittled the problem. The result, considered in the next sections, was the production by the state of a very fragmented air pollution and environmental policy that targeted urban crises, and the proliferation of civil procedures arising from disputes between private individuals.

Social responses to environmental problems were mild during the 1960s and 1970s. The first tragic consequences of unregulated industrialization and urbanization were just becoming visible. Disasters, such as the flooding of museums in Florence in 1966, the gradual sinking of Venice, the landslide in Agrigento on 19 July 1966, all happened in the span of a few years. Most Italians were shocked but remained indifferent to these events: choosing to ignore the magnitude of the problem, they interpreted the disasters as isolated unrelated phenomena.

The first exception to this general state of ignorance or apathy were groups of intellectuals and architects that formed to tackle the most urgent problems and produce immediate solutions. These groups tended to interpret the disasters as symptoms of a larger process in the economic development of the period: unregulated industrial growth, heavy immigration from rural to urban areas, abuses of land and natural resources.[12] In most cases, their "mobilization encouraged policies to increase city planning for historic sites" rather than policies for the protection of the environment (Reich 1984, 388). Proposals included new city plans capable of integrating safely immigrants or proletariats into urban areas and of protecting historic sites, lists of specific policy measures for government intervention, and exhibits to raise public awareness.[13]

The second exception were the reactions of conservation groups, such as Italia Nostra and the Istituto Nazionale di Urbanistica.[14] Concerned with the preservation of the cultural and natural heritage of the country, they too favored, at least during this period, the protection of urban and artistic areas from degradation. They published periodicals, distributed fliers, and organized conferences on the effect of environmental degradation on artistic sites.[15] They reached private individuals, elementary schools, employers' organizations, and public authorities with alarming statistics of crises in urban centers.

Both kinds of social movements could not, and did not, pressure the state to adopt a national policy for the environment.[16] First, their interests in urban areas and culture did not ask, with some exceptions aside, for an environmental policy proper.[17] Second, they lacked the organizational strength to affect change. Until the mid-1980s, these movements boasted small memberships and suffered from very limited access to national media structures[18] and the political system.[19] As Reich noted in the case of architects and intellectuals, the movement lacked the resources to make the environment a major field of political competition" (1984, 389). Their requests were notoriously ignored or only partially considered once in the hands of government bureaucrats (Rovelli 1988, 90–91).

Political parties remained essentially indifferent to the environmental crisis of the 1960s and 1970s (Dente and Lewanski 1983, 112; Benedizione, 1988). The Communist Party (PCI) was at the time the major opposition party to the Christian Democratic government.[20] Its energies were predominantly centered around industrial conflicts and workers' rights. In the party discourse, pollution and the protection of the environment were mentioned mostly in the context of protecting workers' rights at the workplace. Although it criticized the intellectual movement and Italia Nostra as bourgeois and aimed at the preservation of the capitalist mode of production, the PCI did not have an alternative, comprehensive view. In 1971, a party-affiliated research institute, the Gramsci Institute, organized a conference to place "environmental problems within the ideological boundaries of the party as part of the class struggle and workers movement" (Reich 1984, 390). Not until 1979 did the PCI establish a League for the Environment within its organization and join other parties and independent groups to start articulating an ecological policy and to recognize, if not incorporate, the ecological movement. As late as 1990, the leader of the party, Occhetto, would state that the "horizon of the Left must go beyond the politics of class to embrace new issues such as the relationship . . . between man and nature."[21]

For the Christian Democratic Party (DC), in power since World War II, producing a large scale, coordinated response to the environmental crisis would have been impossible. The DC was a highly fractured party with a well-established clientele associated with each faction. For decades, the party as a whole retained control over the government and the administration through the electoral support that clientele groups gave it when supporting their respective DC factions. The established clientele included industrialists, bankers, professional organizations, the Church, and, possibly, even the mafia; it did not include environmental groups. The absence of a large-scale social movement that could penetrate the DC organization made it impossible for the party to create a consensus that would initiate a planned national policy. Accordingly, while the DC had mobilized in favor of some clientele groups, it did very little to articulate a policy to resolve the environmental crisis at hand. It was satisfied with initiating or supporting emergency initiatives to solve the most severe crises.[22]

Starting from the mid-1980s, political parties turned seriously to the issue of environmental protection. Large-scale social movements also emerged.[23] These changes would account for a number of major legal achievements and the belated transposition of several Community directives. As far as the 1960s and 1970s were concerned, however, the prevalent mood of apathy yielded a body of disjointed and very limited laws and administrative structures focused primarily on the protection of urban areas, where the disasters had had the most obvious impact,[24] and an increasing regulation of polluting activities by courts, as these settled disputes between private actors. The next sections describe in detail these legal and administrative developments.

Fragmented National Law and the Recourse to Civil Law

As of 1980, Italy had fewer laws protecting forests, soil, water, air, and mountains than most European countries. It also had the fewest public agencies for the environment, research institutes, state university courses and chairs in ecology in Europe (Cederna 1975). Italy could only boast an "incomplete . . . contradictory . . . excessively fragmented" approach to the environment (Bettini 1984, 149).[25]

Specifically, the legal context awaiting the Air Pollution Directive included a number of emergency laws designed to resolve immediate environmental threats to urban areas, and a growing number of cases between private individuals or groups, solved with reference to the Civil Code, regulating pollution in a case-by-case fashion in those areas not covered by the emergency laws.

In 1980 there existed only two major laws protecting the environment, and a number of minor special laws.[26] The major laws were the Air Pollution Law of 1966 (no. 615), considered in detail here given its importance for the SSD, and the Water Pollution Law of 1976 (no. 319).[27] No other major law dealt with pollution in other sectors of the environment, such as waste management, control of chemicals and noise.

Law 615 on air pollution was primarily an emergency measure taken to "reduce levels of sulphur dioxide (and particulates) in those areas of the country where ground level of such pollutants was extremely heavy at the beginning of the sixties" (Dente and Lewanski 1983, 108). As such, it was designed for areas most afflicted by SO_2 and smoke pollution and targeted with fairly precise terms major emitters of those substances.[28] The law said nothing about air quality standards (only about emissions) and little about the remainder of the national territory, other types of pollutants, or future objectives.[29]

Specifically, the law applied only to special zones of the country selected in terms of size of cities, climatological conditions and their location in either Northern or Southern Italy. In the North, small cities, with a population of 70,000 to 300,000, with particularly unfavorable industrial, climatological or urban conditions and cities with over 300,000 inhabitants would be covered. In the South, cities, with a population of 300,000 to 1,000,000, with particularly unfavorable conditions and larger cities, exceeding 1,000,000 inhabitants, would be covered. In addition, the law applied to areas and towns that needed protection for reasons of public interest or extreme pollution.

Law 615 targeted only the most powerful emitters: industrial plants, thermal installations, and motor vehicles, especially diesel-powered vehicles.[30] It gave each type a series of specific instructions. For industrial plants, instructions required observance of the 'strictest limits allowed by technology,' followed later by specific quantitative emission limits,[31] requirements for antipollution equipment, and treatment before discharge. For thermal installations, these included specific dimensions of chimneys and operating licenses, issued by the provincial work inspectorates, for large installations, and methods for discharge.[32] For diesel-powered vehicles, these included limits on the opacity of exhaust emitted.

Law 615 thus applied selectively to areas in most need by targeting only the worst pollution problems in *urban areas* only. The law did not regulate other geographical areas or smaller polluters. Most importantly perhaps it did not specify *standards for air quality*. This would have required a far more comprehensive approach: the identification of desired standards for the entire territory, a strategy for attaining those standards,[33] and the creation of administrative units to monitor pollution levels throughout the territory and compliance.

After 1966, a few special laws responded "to particular contingencies and emergency situations" in the field of air pollution (Guttieres and Ruffolo 1982, 69). Among the most notable was Law 171 of 1973, known as the Law for Venice: it applied the principles limiting air pollution from motor vehicles of Law 615 to barges on the Venetian lagoon. Law 437 of 1971 specified carbon emission limits for automotive vehicles with positive ignition. Law 584 of 1975 prohibited smoking in certain public areas, such as cinemas, schools, and theaters, and on public transportation. With the few exceptions of additional minor special laws, in 1980 no other national legal measure existed in the area of air pollution.[34]

This dearth of national laws might have been justified by the presence of regions with legislative powers of their own.[35] The delegation of powers to regions, partly found in the Constitution of 1948 and in DPR 616 of 1977, was designed to relieve the state from drafting and applying national laws. In the Italian case, though, the state explicitly retained responsibility for coordinating the activities of regions with broader guidelines, legal and administrative, informed by a general agenda.[36] The state produced no such guidelines for the environment with the exception of course of special emergency laws;[37] there was "no adequate coordination action on the part of the state organization corresponding with this fragmentary regional action, in order to ensure a single cohesive policy," neither at the time of delegation nor later, when it became clear that some regions had developed significant programs while others had not (Guttieres and Ruffolo 1982, 190).[38] One of the few ecologists to work in the government would still urge the state in 1984 to "formulate norms for regions to follow . . . to provide a legal framework" and complained about the "lack of coordination" produced by "a law here and a law there" (Accademia Nazionale dei Lincei 1985, 6). Italy had no policy for the protection of its air and environment.

In a country seriously affected by pollution, this situation naturally led a number of individuals and groups to resort to alternative methods of safeguarding their interests. At the time, experts had written:

> The incomplete nature of the preventive programme and of the overall management of the environment have necessitated the development of an instrument of protection for injured interests. (Guttieres and Ruffolo 1982, 183)

More specifically:

> In a situation of general ineffectiveness of the regulatory activities carried out by the national or local administration, the courts in the sixties and seventies gradually became a very important actor in the protection of . . . interests. (Dente and Lewanski 1983, 117)

Article 884 of the Civil Code, prohibiting intolerable emissions of noise, gas, vibration, and smoke on the property of others, became an especially popular remedy. The article was theoretically a protective instrument only in the case of neighbors and safeguarded primarily their property, rather than their health. It was increasingly used by private individuals and parties against the polluting activities of industrial or other types of plants.[39] Article 2043 of the Civil Code, protecting from unjust damage, was similarly utilized. Gradually tribunal judges learned to accept and even support the application of these principles as a way of handling the problems of the environment.[40]

This use of civil law framed environmental problems as problems for private interests, rather than public interest. It regulated pollution by upholding principles, usually related to individual rights, very different from those of Law 615. It was found to be "of necessity intermittent in nature, linked to the reasoning behind each individual case and to the discretion of the individual judge involved" (Guttieres and Ruffolo 1982, 183). It relied on tribunals, traditional law, and private initiative rather than an administrative public structure designed to ensure the protection of the public good. While the system may have been a valid tool to safeguard the environment in numerous instances, it was not part of a national policy designed by the state. It grew next to the legal system and posed an additional challenge to any future effort toward a coordinated national policy.

The Absence of a National Administrative Apparatus

A second consequence of the weak response of social and political actors to the environmental crisis was a poorly developed administrative structure. The national administrative context awaiting the Air Pollution Directive was fragmented and disjointed, reflecting the poor legal body it was intended to actualize. There was no national administrative unit, like a ministry, solely dedicated to the protection of the environment and capable of applying the directive. Law 615 had failed to create a national administrative structure for air quality protection.[41] By the early 1970s, the Italian government was spending between 15% and 20% of what other member countries went spending on the administration of environmental laws.[42]

The division of responsibilities occurred among at least thirteen different ministries. Table 5.1 lists the ministries and their responsibilities. For most ministries, table 5.1 shows a reasonable assignment of responsibilities. Functions only at times overlapped across ministries, such as those of the Ministries of Public Works and Education in the areas of countryside and territory. The absence of any higher coordinating body, or of a ministry that united all of these functions into its own hands, was however obvious, as lamented in a conference in 1984 by Giuseppe Montaletti, the only expert on ecology in the Ministry of Cultural and Natural Assets.[43] The division of responsibilities had occurred over time as special laws were drafted and administrative support was needed. In each case, the state selected the most appropriate ministry or ministries for the problem at hand. Then, as new problems arose, the various interested ministries were summoned to respond.

Table 5.1. Division of Responsibilities for the Protection of the Environment in Italy at the National Level in 1980

Ministry	*Responsibilities*
1. Public Works	Public waters, town planning, organization of territory
2. Industry	Exploitation of resources, control of production and use of hydrocarbons
3. Transport and Civil Aviation	Atmospheric noise and air pollution from engines
4. Mercantile Marine	Protection of maritime areas
5. Agriculture and Forestry	Protection of forests and control of production activities in agriculture
6. Budget and Economic Planning	Distribution of financial resources
7. Interior	Control over certain sources of pollution (such as heating installations) and peripheral organizations for state intervention (such as fire service)
8. Labor	Protection of safety and quality of the workplace
9. Education	Protection of the countryside, natural beauties
10. Scientific Research	Coordination and financing of research and studies
11. Justice	Studies on legislation and the activities of the judges
12. Cultural and Natural Assets	Protection of cultural heritage
13. Health	Protection of health of the general population, and of air

Source: Data from Mario Guttieres and Ugo Ruffolo, *The Law and Practice Relating to Pollution Control in Italy* (London: Graham and Trotman, 1982).

There were more than thirty auxiliary technical and public bodies and parliamentary organizations active on the environment, created as necessity demanded, with mostly advisory functions and often burdened with additional unrelated functions.[44] They included the National Research Council, with experts on scientific matters and the Laboratory for Atmospheric Pollution, conducting research on air quality (Guttieres and Ruffolo 1982, 22). Two parliamentary commissions on ecological problems, under the Senate, and on water problems, under the Chamber of Deputies, supplied opinions and data on pollution (Guttieres and Ruffolo 1982, 23). Three councils also worked within ministries or in auxiliary positions as consultants to central administrative bodies: the Superior Councils of Health, Public Works, and Antiquities and Fine Arts.

Only a few, indecisive efforts were made to create a superior coordinating body, such as a Ministry of Environment. Law 615 of 1966 created only an advisory Central Commission Against Atmospheric Pollution, located in the Ministry of Health and with representatives from interested ministries. In July 1973, Prime Minister Mariano Rumor instituted an environmental ministry without portfolio, that is, lacking its own budget and administrative staff. Then, in March 1974, Rumor, heading a new government, merged the ministry with the Office of Cultural Assets (later Ministry of Cultural and Natural Assets), and still deprived it of its own portfolio (Cederna 1975, 6n1). The tendency to ignore ecologic matters was reconfirmed later that year. In December 1974, Prime Minister Aldo Moro, at the head of yet another government, issued a decree-law, generally acceptable only in situations of emergency when originating from the executive branch and to be approved within sixty days by both houses to become law, for the creation of a ministry on the environment and cultural heritage. The Senate, in a very crucial decision, amended the decree, accepting the ministry but "granting real powers only for activities on cultural assets" (Reich 1984, 386).[45] The ministry included around ninety experts on artistic and administrative matters and only one on ecology. As Giuseppe Montaletti, the expert on ecology, argued in 1984: "I have been the only representative of nature . . . a single voice and often a *vox clamantis in deserto*."[46] In essence, the state committed itself only to the cultural heritage endangered by the environmental crisis rather than to the environmental crisis itself.[47]

A well-coordinated and powerful network of regional, or even local, administrative structures may have explained the absence of a national structure dedicated to the environment. There was, however, no such network. Both major laws, on air and water pollution,[48] assigned to regions a number of important administrative functions but delivered only some financial support.[49] Some regions developed structures, but most remained essentially passive. No national agency sought to remedy the problem in the following years. DPR 616 of 1977 formally handed much of the responsibility for the protection of the environment to regions. It did not specify long-term targets, strategies, deadlines, and costs of noncompliance.[50] Most regions developed different, at times competing independent administrative units primarily to cope with emergency situations.[51]

Specifically in the field of air pollution, Law 615 of 1966 had assigned to regions[52] the duty of ensuring its implementation. The duty consisted of three parts: stations' erection and data collection, planning a regional administrative apparatus,[53] and intervention to stop polluters from exceeding limits.

For the first assignment, the state simply failed to deliver the promised financial support for regions necessary to build monitoring stations (Capria 1992, 145; 1991, 69). For planning and intervention, Regional Committees for Air Pollution were instituted. These committees were composed of members of the national, regional, and provincial administrations, representatives of regional economic interests, and representatives of associations of local authorities (Dente and Lewanski 1983, 108–9).

Planning and intervention were however compromised from the start by the lack of information that should have originated from monitoring stations. Some committees, especially those in more industrialized regions and under serious environmental threat, improvised and created on their own data collection systems and planned individualized intervention strategies. These committees met with higher frequencies, involved more participants, relied at times on outside help for data collection,[54] and pursued disjointed policies in response to their needs.[55] Most regions though remained passive.

It is interesting to note that, as courts became involved in the regulation of disputes between private individuals, they too naturally developed a sort of administrative apparatus. The Territorial Ecological Group, in the Computer Center of the Court of Cassation, provided experts' evaluations and information on past cases (Guttieres and Ruffolo 1982, 23). By and large, however, Italy did not have an administrative apparatus capable of implementing the Air Pollution Directive.

Explaining Implementation

The Air Pollution Directive asked Italy to develop a comprehensive policy for the protection of air. It specified air-quality standards, regardless of current levels in individual countries, applicable throughout the entire national territory as well as future values. It asked for the creation of an administrative apparatus capable of disseminating information, monitoring compliance, and applying sanctions throughout the country. The directive thus asked for a policy in which strategies for the attainment of objectives would be spelled out: How to measure current levels and how to gauge the gap from the quantities specified by the directive? Who should be targeted, according to which criteria, and to what extent? Who should be in charge of this selection and application?

Law 615 of 1966 had asked for much less. As an emergency measure, it had targeted only the largest polluters and only in specific areas. Given the absence of information concerning levels of pollution, and the difficulties involved with establishing standards, the law merely *restrained*, with vague terms, polluters. Administratively, the law did not create and thus could not rely on a ministry devoted to the environment. The attempt to delegate to regions the duty of application proved to be a mere dismissal of responsibility: the state did not deliver the promised financial support or guidance.

The growing body of civil law cases, resolved by the Court of Cassation, represented a case-by-case approach that situated the regulation of the environment into the sphere of private interests. The presence of these civil cases deprived the discourse on the environment of that basis that is the ideological foundation of most state policies: the importance of the subject at hand, in this case the environment, for the collective well-being and long-term interest of the nation. Essentially impossible to dismiss, this legacy stood as an additional challenge to the development of a public policy on the environment.

The state thus faced a request for the formulation of a policy that, however important, had been ignored publicly and addressed in erratic fashion privately. In response, the state produced the DPCM of 1983, a belated decree that, rather than creating new structures or laws, *reflected* the legacy of the preceding decades. It would have been impossible for the Italian state to gather the consensus and resources necessary for the directive after decades of apathy. With the exceptions of the correct limits of SO_2, every other measure that *specified limits* was openly ignored: limits for smoke, winter median values, peak values, and the fixed relationship between smoke and SO_2 values. Measures that required planning for the future were also ignored: guide values and the standstill principle. The 1983 DPCM left untouched the weak administrative status quo set up by Law 615. The decree was very vague in the delegation of responsibilities, imposed no serious penalties, did not create a ministry for application or centers for research.

The policy legacy of the Italian state also explains the poor application of the 1988 DPR. The financial resources and political commitment needed to subsidize regions, to establish monitoring stations, and to design plans for action were clearly missing. When only a handful of regions presented plans after the 180 days allowed, the Ministry of Environment did not solicit or punish the noncompliant regions.[56] These failures reflected a pattern from the past.

The partial, proper transposition of the DPR remains to be explained. One should note, in addition, that the DPR, unlike the DPCM of 1983, did spell out with unusual clarity the delegation of duties to regions and was produced with the collaboration of a new Ministry of Environment, whose creation dated to 1986. The application measures were more promising than those of the DPCM. To account for these aspects of the DPR, the attention must turn to changes in the policy legacy and interest groups that had taken place since the mid-1980s.

Evidence points to a transformation in society, political parties, and the legal and administrative state structures in favor of the protection of the environment. From the mid-1980s on, large-scale social movements formed to protest plans to install new missiles for nuclear war and the construction of nuclear power stations,[57] traditional political parties listed an environmental policy in their agendas, and new electoral lists and a Green Party formed specifically for the protection of the environment.[58] Memberships in green organizations exploded: the WWF reported an increase from 30,000 members in 1980 to 274,000 in 1990, and the Lega Ambiente reported an increase from 15,000 members in 1983 to 50,000 in 1990 (Davidson 1991, 48).[59] The Italian government, now with representatives of the Green movement, began producing major laws and administrative structures, such as the Ministry for the Environment in 1986 (endowed with a portfolio[60] and a full time staff),[61] a major law on the protection of the countryside,[62] the creation of eight new national parks (after fifty years of inactivity), and a special effort to deal with the problem of 'forgotten' Community directives.

These were changes that transformed the national policy legacy on the environment. They were overdue domestic reactions to an intolerable situation. They explain why the Italian state drafted the DPR of 1988:[63] the decree was one of many legal initiatives of the time that tried to overcome the past. It reflected a concerted effort to regulate pollution of waters, land, and air.

One quality of this internal transformation explains the applicative limitations of the decree. The period from 1985 to 1990 witnessed the construction of an administrative apparatus, but not its proper functioning. Laws were made with relative ease. Administrative building was beginning to be underway but could not ensure the immediate application of new laws. As Law 615 of 1966 did, the decree chose to rely on regions but failed to deliver to them the necessary logistical and financial support.

In sum, the implementation of the Air Pollution Directive, whether in the early 1980s or thereafter, can only be understood in light of the policy legacy on environmental protection and air quality in existence in Italy at those times. The absence of a serious mobilization on the part of social and political actors to the environmental crisis of the 1950s and 1960s explains the absence of a national environmental policy. In 1980, the legal body on environment boasted a number of unrelated emergency measures, with only two major laws. One major law, Law 615 of 1966, dealt with the frightening quality of air by targeting major polluters in urban centers, that is, those areas most obviously threatened by the crisis and naturally the target of the few social groups that had mobilized. A negligent state had taken no real measures to ensure the application of the law. There was no administrative structure capable of applying a major law on air quality. The directive was accordingly poorly transposed and applied with a decree, the 1983 DPCM, which clearly omitted many of the important requirements of the directive and took no real administrative measures.

An almost inevitable change in the Italian approach, at the level of social and political actors, to the environmental crisis occurred in the mid-1980s and explains the production of a second decree, the DPR of May 1988. The DPR, although years after the original deadline for compliance, specified faithfully the requirements of the directive and applicative measures. The transformation had just started, and Italy was still unprepared to ensure a proper application of the law. The upgrading of the administrative structure would follow years after that of the legal structure.

6

The British Case

Attempted Distortion of a Legacy

Great Britain refused to transpose fully the SSD into legal form until 1988. For eight years, it relied on its Clean Air Act of 1956, three additional minor acts, and the Alkali Act of 1874 as official translations. These acts specified only one of the requirements put forth by the directive, namely the construction of a monitoring network. Application, too, was inappropriate, although certainly better than transposition. Some of the most important air quality standards were met in time in spite of poor transposition. Other requirements, such as the production of plans for complying with future guide values, were not met until many years after 1980.

In this chapter, I explain Great Britain's late transposition and partial timely application in terms of the preexisting legal and administrative state policy on air pollution. In 1980, five or more legal entities regulated various aspects of air pollution. These entities focused on *emissions* rather than air-quality standards and imposed no national standards for those emissions. The directive unrealistically asked for the introduction of fixed air-quality standards for the whole of Great Britain: for the most part it was not, therefore, transposed. The preexistent legal entities had ensured, however, very low concentrations of pollution in several parts of the country and the creation of a national network of stations. As a result, two of the directive's requirements were anyway applied. The other requirements proved far more demanding for Great Britain and were, consequently, not properly applied.

The directive's demands on the organization of interest groups engendered dynamics that were only secondary for implementation. As in Italy, conservation groups were too weak to mobilize in favor of the directive. Producers of pollution, potentially hurt by the directive, approved the state's reaction to the directive and allowed it to protect them.

Time and Extent of Transposition and Application

The Department of the Environment warned confidently in 1976 that the United Kingdom would attempt to meet the incoming requirements of the Air Pollution Directive by making use of existing legislation. In an explanatory memorandum to Parliament submitted that year, the department had written: "It is thought that if the Council adopted this Directive it could be implemented in the UK without further legislation."[1] Until 1986, that was indeed the approach followed by the government to implement the SSD.[2]

The department wrote to the European Commission in 1982 that the legislation used for transposition would include the existing Clean Air Act of 1956 and its amended version of 1968, the Alkali Act of 1874, and the Control of Pollution Act of 1974.[3] The department also relied on circular 11/81, issued on 27 March 1981.[4]

As such, the acts of 1956, 1968, 1874 and 1974 did not include the most important principles put forth by the directive. None contained any reference to the required limit values for concentrations of SO_2 and smoke, compliance plans for exceptions, the standstill principle for SO_2 and smoke, or the guide values for SO_2 and smoke. The acts were concerned with eliminating, in the case of smoke, or controlling, in the case of SO_2, emissions. They targeted producers, and not standards for air quality.

The Clean Air Act did require the construction of monitoring stations capable of measuring concentrations of SO_2 and smoke in the air at ground levels. The act therefore faithfully translated the directive's requirement for monitoring stations. Actually, by 1980 a number of stations were being dismantled. Circular 11/81 requested, however, that the long term plan of reducing the stations be reconsidered in light of the requirements of the directive.[5]

Circular 11/81 could not be considered an acceptable transposition of the directive's remaining requirements. First, the circular was not a national law. All of its requirements, whatever they may have been, did not amount to being legal translations. Second, the circular made compliance with the directive optional anyway: local authorities were asked "to take action thought likely to be necessary" to implement the directive. The circular did not order local authorities to comply with the set standards. As would be expected, in 1988, the Commission issued a reasoned opinion criticizing the use of administrative measures (circulars) for setting air quality standards.

The application of the directive was partly faithful and partly on time.[6] Thanks to the existing emission laws, the most important objectives, the desired concentrations of SO_2 and smoke, were met throughout most of the territory.[7] By around 1980, approximately 5,500 local authorities had issued Smoke Control Orders affecting over 8 million premises (Hughes 1992, 336).

Between 1950 and 1980, industrial SO_2 emissions decreased by at least 50 percent, while power station's emissions at ground level were almost zero. By 1985, most urban and rural concentrations for SO_2 were either free of smoke or well below the limits. On average, urban concentrations stood at 38 and rural concentrations at 28 micrograms per cubic meter.[8]

Monitoring stations existed and functioned well. In a 1983 letter to the Commission, the Department of Environment assured that the network would be kept functional in spite of plans, made before 1980, to reduce it:

> The UK has had an extensive network of monitoring stations for smoke and sulphur dioxide for many years and measurements will continue in all areas where there is a possibility of approaching or exceeding the limit values. . . . Powers exist to require measurements to be taken wherever necessary. (As reported in Haigh 1990a, 189)

Areas in breach of the limits concerned smoke alone and were closely monitored with existing stations built under the Clean Air Act of 1956, numbering around 400.[9] Reports of current concentration levels were forwarded on time to the Commission on a regular, yearly basis (Haigh and Hewett 1991, 83).

The British government did not, however, draft plans for improvement for those areas in breach of the directive, nor really take provisions to ensure guide values for the future. Sanctions for noncomplying regions were nonexistent. In 1985 the Commission informed the Council that Britain had not submitted the necessary plans for improvement of specific areas.[10] The Commission subsequently met with representatives of the Department of Environment, but the complaints stood firmly. In the reasoned opinion of January 1988, the Commission argued that improvement programs had not been carried out (Haigh and Hewett 1991, 84).

Starting in 1986, the Department of Environment and the government began to realize that a stronger, unified legal measure would be necessary if the legal and administrative requirements of the directive were to be met. This measure would set air quality standards, reinforce monitoring networks, and centralize most administrative and legal functions in the hands of a central government's figure. In 1988 and 1989, the British government passed the Air Quality Standards Regulation which transposed fully, however late, the SSD's requirements.[11] In 1990, the government passed the Environmental Protection Act.[12] A major legal advance, the act conferred to the secretary of state extensive power to make administrative regulations in order to meet EC obligations. Specifically, Section 3 empowered the secretary to establish "standards, objectives, or requirements in relation to particular prescribed processes or particular substances." Thus, the secretary could now "prescribe standard limits for the concentration or amount, generally or in a particular time period, of a substance released from a prescribed process into any environmental medium (air, land, water)" (Hughes 1992, 322). Section 3(5) also empowered the secretary to design plans for progressive tightening of air quality standards. Thus, administrative functions were removed from local authorities and centralized into the secretary of state. Sanctions for breaching EC air quality standards were made heavy, ranging from £20,000 or more to two years of imprisonment, and could apply to individuals even when acting on behalf of companies or collectivities. The secretary made use of these new powers, and application eventually took place throughout most of the territory.

Why did the British government rely for many years on a circular and preexisting laws to comply with the directive? Why did Great Britain resist implementation until 1988 and 1990?

The Legal Approach to Smoke before 1980: Control of Emissions

Britain, prior to 1980, possessed an effective legal apparatus for combating smoke: the Clean Air Act of 1956, aimed at regulating *emissions* rather than at ensuring prescribed concentration levels (as the directive asked). SO_2 pollution was also approached in terms of emissions. The Alkali Act of 1874 charged an inspectorate with regulating *emissions* at industrial sites and, in theory, at power stations. Administratively, both acts relied on decentralized, informally operating structures. The directive asked, with its focus on air quality standards, for a reorientation of this emissions-oriented legal approach, and for the uniform application, by a centralized administration, of the new standards. The gap between the directive and the existing legal and administrative context was huge and hardly surmountable. Legally, there would be no formal compliance with the directive until 1988 and 1990. Administratively, the decentralized system used for emissions was charged with complying with the directive. At the time, it could comply with some, but not all, of the requirements. This section analyzes the legal approach to smoke pollution used in Britain prior to the directive.

For decades before 1980, the British government had faced extremely high levels of air pollution in its urban and, to a lesser extent, in its rural areas. Serious incidents of atmospheric pollution occurred in London in the late 1890s, 1921, and 1948, in Manchester in the 1930s, and in other industrial and heavily populated cities before World War II and soon after it.[13] The most serious and influential incident occurred in London in 1952, between 5 and 9 December, when a thick fog composed mostly of smoke and sulphur dioxide settled into the town and temperatures did not rise above freezing point. The conditions were alarming:

> At times visibility was below ten yards, and for substantial periods it
> was below twenty yards. Fog invaded public buildings and on the

night of 8 December the audience with balcony seats in the Royal
Festival Hall could not see the stage. (Clapp 1994, 44)

Most alarming was the increase in death tolls during those five days. Around
four thousand deaths were a direct result of the poor air. This number was
only surpassed in the modern period by the outbreak of influenza in 1918.
The quantity of pollutants was ten times the normal figure for that time of
year.[14]

The reaction of the government was immediate.[15] The Ministry of
Health set up internal inquiry committees. Parliamentary requests for a public
committee led to the formation of an inquiry chaired by an industrialist, Sir
Hugh Beaver, in May 1953. "Working with unusual speed" the Beaver
Committee produced an interim report after six months and a final report a
year later (Clapp 1994, 49).

The most important characteristic of the report was its focus on *emissions
of smoke*. In this, the report fully reflected Britain's historical approach to
pollution law.[16] The committee calculated the costs of unabated smoke for the
population and called for a reduction in emissions by eighty percent within
fifteen years.[17] The resulting Clean Air Act of 1956 accordingly prohibited
emissions of dark smoke from a chimney of 'any building' including domestic
fires,[18] specified that individuals would be held responsible for breaches,[19] and
listed a number of extenuating circumstances under which violations would
not be punished.[20]

The act made local authorities responsible for ensuring that no area
produce black smoke. Local authorities were also allowed to deem certain
areas exempt and others 'control areas' (Clapp 1994, 50).[21] In such 'control
areas,' emissions of any smoke would be considered an offense (Hughes 1992,
334). National inspectors, from the Alkali Inspectorate, were charged with
overseeing industrial smoke emissions for steel and iron, and smoke emissions
from power stations.[22]

The Historical Roots of Smoke Emission Controls

The Clean Air Act of 1956 focused on emission controls for smoke for important reasons. The act was the last event in a century of legal struggles all
targeted at emissions.[23] Between the 1840s and the 1950s, individuals and
organizations had mobilized mostly without success to convince the legislature that emission controls over all smoke-producing sites were the necessary
solution to air pollution in the country. The merits of the Clean Air Act rested
in its precise definition of smoke,[24] its inclusion of domestic fireplaces, and its
clear quantitative focus on emissions. It was in this sense a victory belonging
fully to an historical context.

The act amended fundamentally the Public Health Act of 1875.[25] No bill designed to introduce similar changes to the 1875 act had been passed in Parliament between 1875 and 1955. The Public Health Act of 1875, in Section 91, controlled smoke emissions with vague measures:

> Any fireplace or furnace which does not as far as *practicable* consume the smoke arising from the combustible used therein . . . and any chimney (*not being the chimney of a private dwelling house*) sending forth *black* smoke in such quantity as to be a nuisance, shall be deemed to be a nuisance.[26] (Italics added)

No specification as to the meaning of 'practicable means' and 'black smoke' were offered. Domestic fires were exempted. Enforcement was difficult. Further increasing difficulty of enforcement, the 1875 act spelled out that the Court

> shall hold that no nuisance is created within the meaning of this Act, if it is satisfied that such fireplace or furnace is constructed in such manner as to consume as far as *practicable*, having regard to the nature of the manufacture or trade, all smoke arising therefrom, and that such fireplace has been carefully attended to by the person having the charge thereof. (Italics added)

All subsequent proposals attacked one, some or all of the 1875 act's limitations.

The Fog and Smoke Committee was the first major organization to mobilize against the existing legal status quo.[27] All of its initiatives focused on controlling emissions more fiercely. The committee appointed a subcommittee to investigate the state of the art in fuel technology, asked an expert to report on the current laws on smoke prevention, organized an exhibition on equipment and design on smoke prevention for public information,[28] and organized meetings with factory owners to discuss strategies for abating smoke production (Ashby and Anderson 1977a, 13). On 25 November 1880, at a very well-attended meeting in London, resolutions were passed endorsing the committee's intentions to search for alternative methods of combustion.[29]

The committee timidly approached the issue of domestic fires at a second meeting. Members of the Royal Society were given a chance to express their views: "it would not be wise," stated G. J. Shaw, at that 1881 meeting, "to interfere" with the domestic sphere with "any legislation" (The Times, 10 January 1881). The meeting ended with a decision that much research on smokeless fuels and appliances applicable to households was still needed before one could think of "any application for amendment of the existing Smoke Acts, or for new legislation in regard to smoke from dwelling houses" (The Times, 10 January 1881).

The Fog and Smoke Committee was an informal association. In 1882, it was incorporated into the National Smoke Abatement Institution (NSAI). The mission of the NSAI remained very similar to that of its predecessor: search for alternative fuels, tighter sanctions, and control of domestic fires. At the annual 1883 meeting,[30] the following resolution was adopted:

> The period has now arrived at which systematic inquiry is desirable into the application of resources of technical science for the abatement of smoke now largely produced in industrial processes and the heating of houses, as well as into the operation of the existing laws for smoke abatement; and that the Council of the National Smoke Abatement Institution be requested to urge upon the Government the desirability of appointing a Royal Commission for the purpose. (*The Times*, 18 July 1883)

The NSAI retained throughout its existence a careful approach to air pollution, one not shared by Lord Stratheden and Campbell, an "enthusiast for clean air" (Ashby and Anderson 1977a, 16). While the NSAI studied matters, Lord Stratheden and Campbell presented to the House of Lords a total of ten unsuccessful bills between 1884 and 1892. His bills attacked, depending on the political and social climate, different limitations of the current laws. Some bills received second readings; others were immediately attacked viciously and never considered again. Lord Stratheden and Campbell would come after each defeat with solutions to the problems identified in the House of Lords.

The content of each bill inevitably focused on either strengthening emission controls or extending controls to domestic fires. The first bill in 1884 naively asked for the prohibition or regulation of "the emission of smoke from any building." The fourth bill in 1887 was backed by the NSAI. This time its direct attacks on domestic fires eventually evoked angry responses in the House of Lords. The intrusion of spies into private homes and the administrative costs of enforcement were particularly salient topics.

In his sixth attempt in 1889, Lord Stratheden and Campbell focused on three simple ways in which smoke in private houses could be checked. In his seventh attempt in 1889, he asked that authority to enforce the Public Health Act be removed from local governing bodies and be given in part to the central government. In his last attempt in 1892, he proposed alternative ways of controlling emissions, including the adoption of a mix of coke and anthracite, instead of modifications to existing gates (Ashby and Anderson 1977a, 21).

The Coal and Smoke Abatement Society (CSAS), formed in 1899 by the London County Council because of its dissatisfaction with the enforcement of the Public Health Act of 1875 as it applied to London,[31] proceeded to fight the same battle after Lord Stratheden and Campbell's death. The CSAS pressured the London City Council to introduce bills for strengthening current legislation. Changes included the removal of the word 'black' for qualifying smoke, fewer exemptions, the centralization of power, and funding for research and education. Objections again mounted, this time from thirteen different sources, including the Electric Supply Companies of London and the Chamber of Commerce (Ashby and Anderson 1977b, 195). The bill eventually received the Royal Assent in August 1910, after all but the most minor of its requests were rejected.

Around the time of the birth of the CSAS, northern smoke-abatement leagues joined to form the Smoke Abatement League of Great Britain. This became the most powerful lobbying group prior to World War I. Their bill, presented in the House of Commons on 30 April 1913, consistently with its predecessors focused on smoke emissions. Specifically, the league asked for any smoke (not only black) to be recognized as a potential nuisance, for stricter penalties, and for the rationalization and strengthening of adminis-trative control.[32] The bill was ordered a second reading and was eventually introduced in the House of Lords in 1914. Progress was halted once it was announced that a committee would be set up by the government to look into the issue.[33]

The committee had to wait for a reappointment by the minister of health, after the war, to resume work. By 1920, the committee had met fifty times, interviewed 150 witnesses, and was working on a major report. The minister of health had asked for a report listing steps, both practical and desirable, to reduce pollution.[34] The final report included two major recommendations: that all manufacturers use the best practical means to prevent production of any smoke, and that the minister of health be empowered to *set fixed standards of emissions* from time to time, either for the entire country or specific areas. The report also touched upon domestic fires by recommending smokeless heating units.

On 10 May 1922, the report was debated in the House of Lords. A Smoke Abatement Bill was then introduced into the House by the Ministry of Health and discussed on 24 July. Consistent with the past, officials in the Ministry of Health had removed the requirements of fixed standards as impracticable. Lord Newton expressed his disappointment by noting that, without fixed standards, very little could be hoped by relying on the first recommendation. The bill's progress was arrested by the coming of a conservative government in October 1922.

This was an ideal time for industrialists, increasingly worried, to voice their concerns for any amendment of existing legislations. The Confederation of British Industry (CBI) lobbied the Ministry of Health.[35] They sent forth Professor Bone, of the Royal College of Science who held the chair of fuel technology, to argue that any idea of amending legislation should be ignored and that any law that might "imperil the industry in this country and absolutely ruin the commercial efficiency of trade" should be sternly opposed. At the same time, the CSAS and some members of the House of Lords, with the support of public opinion, renewed their pressures for new legislation.[36]

Soon after a new abatement bill was introduced. This time a number of exemptions were made for industry, in light of costs and the nature of specific trades. As to fixed standards, no longer would the Ministry of Health have indiscriminate control. Local authorities were charged with the responsibility of setting the standards and subjected to the approval of the ministry. Lord Newton and others complained about the limits of the bill, calling it a 'manufacturers' bill',[37] and its timing, since there were only few moments available to seriously debate it.

Finally, in 1925, after two more attempts in the House of Commons and one in the House of Lords, a bill reached the House of Lords. This bill strengthened the powers of the ministry, granting it the authority to request reports from local authorities and to empower county councils to act on behalf of local authorities that were evading their duties. The new minister of health, however, also obtained the removal of the provisions controlling domestic fires. A fierce fight erupted in the House of Lords over domestic fires. The bill received the royal assent on 15 December. The resulting act amended the 1875 Public Health Act in some ways: it increased penalties and it somewhat centralized administration of the law. On the negative side, however, it failed to offer a better definition of black smoke, and left domestic fires unregulated.

Three decades later, The Clean Air Act of 1956 thus came primarily as a victory over domestic fires. It brought to an end a century long battle between conservatives, concerned with protecting private rights, and those liberal-minded individuals preoccupied with the consequences of a spoiled public good. The act was a victory in a war over controls of emissions. Importantly, it imposed no air quality standards; it instead created a framework that local authorities could use, if they so wished, to control their own polluters. Before 1980, the British legal mind was thus fully alien to those notions of air quality and national standards that the directive required.

The Legal Approach to SO₂ before 1980: Control of Emissions

SO_2 pollution control in Great Britain focused, as in the case of smoke, on emissions and imposed no national air quality, and in this case even emission, standards. There were two major SO_2 producers in the twentieth century: industries and the national power company. Industries were subject to the request of the Alkali Act of 1874 and later revisions that emissions of SO_2 be limited to the greatest possible extent by the use of the 'best practicable means'. The Central Electricity Generating Board (CEGB) was also theoretically under the Alkali Act. Practically, however, it enjoyed remarkable freedom. Its own solution to curbing SO_2 pollution tackled, consistently with other pollution policies, emissions: in the 1960s, chimneys sufficiently tall to emit SO_2 at high altitudes were built to remove all pollution from ground-level air.

Private industries were responsible for approximately 20 percent of SO_2 production in the postwar period. The Alkali Act had regulated their emissions since 1874 by requiring industries to prevent SO_2 emissions with the use of the best practicable means available to them. Other subsequent revisions and alternative acts, up to the Health and Safety of Work, etc. Act of 1974, preserved the principle essentially intact.[38]

'Best' and 'practicable' were left intentionally undefined in the original act and all subsequent revisions. Inspectors were intended to be the central figures in settling production limits. They were instructed to take into account costs of reducing pollution,[39] location, production requirements, topography, prevailing winds, and community needs for each specific industry and factory. Financial considerations for local companies and communities traditionally played an especially central role in the assessment of what a particular factory should or could be expected to do.[40] In the words of an inspectorate, "economics [were] an important part of the word 'practicable' . . . most of our problems [were] chequebook rather than technical."[41] In his 1966 Annual Report, the Chief Alkali Inspector explained how the "expression 'best practicable means' . . . [took] into account economics in all its financial implications," which included "the wider effect on the community."[42]

This conceptually elastic law naturally gave rise to flexible emission standards. Different factories or industries needed to observe different requirements. A standard could be set for a certain industry, but exemptions for specific factories would be granted for adjustments to meet local circumstances. Rather than troubling lawmakers, administrators, or industrialists, these differences were praised by most and increasingly became central features of the British system. The secretary of state for the environment wrote assuredly in 1977:

> Except in clearly defined cases, we believe it is better to maintain gradual progress in improving environment in light of circumstances and needs than to operate through the formation of rigid national emission standards which may be in particular circumstances either unnecessarily harsh or insufficiently restrictive.[43]

A publication by the Department of Environment in 1976 enthusiastically noted how the British pragmatic approach permits the establishment of individual standards for polluting emissions from particular factories which can be made continually more stringent in the light of technical advance and of changing environmental needs. (Department of Environment 1976, 3; as quoted in Vogel 1986, 76)

Industrialists as well approved the principle and, interested in keeping the flexibility alive, genuinely complied with requirements imposed on them. Industrial emissions in fact decreased continuously over the century, even in the 1950s, 1960s and 1970s, times when industrial production greatly increased.[44]

The act, strongly established in industries, should have applied to the CEGB, undoubtedly the largest producer of SO_2 in Britain. With no actual controls on its power stations, in 1955 the CEGB was responsible, by some accounts, for 80 percent of all SO_2 emissions.[45] The company was, at the same time, the only supplier of energy at the national level. Monopoly of power supply, and support from the CBI, freed the CEGB from having to comply with the loosely worded Alkali Act.

For years prior to the Clean Air Act, CEGB's directors and staff had taken decisions over expenditures, investments in alternative energies, research agendas and pollution controls undisturbed by governmental influence or the Department of Environment. The company, in the words of an environmental expert at the Institute for Environmental Policy, made "British air pollution policy."[46] It openly rejected most opportunities to investigate alternative reduction strategies in the 1940s and later, such as flue gas desulphurization (FGD) or alternative methods for treating fuels.[47] After experimenting with desulphurization at two power stations, Fulham and Battersea, before World War II, the CEGB opposed any further government proposal and considered the matter closed.[48]

In 1955, the Beaver Committee, architect of the Clean Air Act and concerned with designing a measure as acceptable as possible, was careful to respect the CEGB's domain. The committee recommended, but did not order, the use of flue gas desulphurization for the CEGB in light of great results in Germany and elsewhere.[49] Otherwise, it left the problem of SO_2 to the Alkali Inspectorate, an agency that everyone knew to be impotent in its relations with the CEGB.

The company's undisturbed hegemony over SO_2 production came to an end in the 1960s as high levels of SO_2 attracted public attention and the media. As public pressures for reform mounted, the CEGB searched for a solution. The eventual answer came in terms of *emissions* regulation. Power stations would be moved away from urban settings into rural areas and chimneys would be lengthened to altitudes that could guarantee almost no

SO_2 presence in ground-level air (Rose 1990, 123). The costs would be cheaper than any manipulation of fuels (such as FGD), use of alternative fuels, imposing limits on production, or setting air standards. The Alkali Inspectorate, theoretically in a position to object, offered no opposition. It would not do so even after the CEGB was practically proven responsible for causing acid rain in Scandinavian countries and Britain itself.[50]

The British legal approach to SO_2 air pollution was thus twofold. The CEGB was free to set its own standards, while the Alkali Act of 1874 controlled private industry. For both the CEGB and private industry, however, the attention was placed on loose standards for emissions.

The Administrative Structure for Controlling Emissions

Unquestionably, the Clean Air Act, the Alkali Acts, and the CEGB's strategy were rather successful. If one considers only production and presence of smoke and SO_2 at ground-level, Britain had clean air by 1980 throughout its territory.[51] For both smoke and SO_2, however, the administrative machinery was remarkably different from that required to apply, and envisioned by, the directive. Application of the directive required administrative centralization, uniform enforcement, and focus on fixed air quality standards. By 1980, however, Britain had a fragmented, informally run, administrative apparatus focused on variable emission standards. Vogel aptly described the British administrative system for pollution as being characterized by an

> absence of statutory standards . . . a flexible enforcement strategy, considerable administrative discretion, decentralized implementation, close cooperation between regulators and regulated. (Vogel 1986, 70)

Application of the Clean Air Act rested primarily with local authorities. Local authorities were given the permission to claim their areas exempt from legal requirements.[52] Attempts to centralize enforcement of control of smoke emissions occurred throughout the twentieth century but systematically failed.[53] The Beaver Committee, which laid the basis for the Clean Air Act of 1956, knew very well that decentralization of control would be necessary.[54] A revision of the Clean Air Act, in 1968, gave the central government the right to require local authorities to declare smoke-free areas, but the right was seldom exercised. In practice, as an expert on the British system put it, the "central government merely gives local authority a framework within which there is considerable scope for creative implementation" (Hill 1983, 15; as quoted in Vogel 1983, 81).

A number of factors, for the most part informally dealt with, guided local authorities in their decision to declare areas smoke-free and to grant exemptions. Vogel identified at least three: the affluence, predominant business, and geographic location of a region.[55] In coal mining regions, for example, miners received as part of their wages coal, intended as a supply for their home heating systems. In these regions, county councils resisted applying the smoke-free law for the obvious reason that the measure would directly forbid miners from using part of their stipend. The National Coal Board had thus indicated its willingness to substitute coal with cash payments. Miners resisted the offer, since the cash would be insufficient to buy the alternative fuels necessary to heat the area once it was declared smoke-free. In other regions with different local dynamics, authorities were able use their powers to declare smoke-free areas. In the cases of SO_2 and the smoke from certain industries, the Alkali Act of 1874 granted alkali inspectors responsibility for controlling emissions. For industrial emissions, the administrative apparatus was strongly decentralized and informal. Decentralization dominated all administrative aspects. Individual alkali inspectors were given full control over individual factories. They were granted the freedom to determine what would be required from each factory in light of all the factors deemed relevant. They determined enforcement procedures, from infringements to sanctions. After imposing limits, each inspector was the "sole judge as to whether the plants and processes under its jurisdiction [were] employing the best practicable means of controlling their emissions."[56]

Informality characterized the system. Mutual trust, rather than diffidence, defined the terms of the game. Only the most flagrant cases, in fact, were brought to court.[57] Differences were otherwise settled informally in most cases.[58] Inspectors chose "to rely on cooperation rather than coercion, working with a particular industry to achieve an improvement . . . over a period of years" (Hughes 1992, 321). Vogel described in great detail the informal and actually exclusive relationship between industry and government that developed over time: private consultations and trust in the good public sense of producers characterized the system (Vogel 1986, 83–84). The 1973 annual report of the Alkali Inspectorate describes the procedure rather well:

> Working parties and discussion groups are set up, consisting of representatives of the industry, its research organization, if any, and the Inspectorate . . . the Inspectorate frequently travels abroad, sometimes in company with representatives, to examine foreign technology. The chief inspector makes the final decision on any standards and other requirements . . . but this only follows mutual discussions with industry representatives.[59]

The collaborative process, based on informality and carried out often with representatives of trade associations, ensured compliance. Participation in the decision-making process by all interested parties led to the formulation of more practicable and enforceable solutions.[60] In its 1974 report, the Royal Commission on Environmental Pollution noted that

> either we have, as now, an authority which because of its close relationship with industry and consequent understanding of the problems is able to assess the technical possibilities for improvement in detail and press for their adoption; or an authority which sees its job as one of imposing demands on industry and which, because of the sense of opposition that approach would create, could not obtain the same co-operation by industry in assessing the problems and devising solutions.[61]

In contrast to private industry, the largest producer of SO_2, the CEGB, was beyond the control of central government or of any emission standards, however informally established. The CEGB's monopoly over electricity supply precluded the development of any administrative structure in its sector. The need to create an administrative apparatus to oversee emissions that could pollute ground-level air was neutralized temporarily by the dispersion of SO_2 at high altitudes.[62]

Next to this decentralized, somewhat informal, and emission-oriented administrative apparatus, stood, somewhat surprisingly, a very well-developed monitoring network. The network, equipped with 1200 stations in 1966 and capable of measuring smoke and SO_2, was among the most sophisticated in the world.[63] Its purpose was to aid smoke and SO_2 authorities in the decision of what emission controls they should impose by providing them with information on concentrations of pollution. The network was largely operated by local authorities, although coordinated by the National Warren Spring Laboratory of the Department of Environment. By 1980, the network was undergoing dismantling. The process was stopped: in light of the SSD, it was the only administrative asset for compliance in the hands of the British government.

By 1980, Great Britain had a decentralized administration for applying the Clean Air Act and a decentralized and informal administration for applying the Alkali Act in private industries. It had no administration for controlling CEGB's emissions of SO_2. The country possessed, at the same time, a powerful network of monitoring stations that could provide data on air quality conditions.

Explaining Implementation

As seen, the Air Pollution Directive asked for a major reorientation of British air pollution policy. Legally, the focus on air quality standards represented a challenge to a tradition built around controlling emissions without imposing air quality standards. Administratively, the directive asked for a centralized, formal, approach: the application by administrators of uniform standards, not to be altered locally, with the added request that regular reports on non-complying regions and polluters be sent to Brussels. This entailed revolutionizing the principles upon which the system had run efficiently for over a century.

Transposing the directive, then, would have entailed committing to principles of air quality and the use of standards contrary to a legacy over a century old. Emission controls, as finalized by the Clean Air Act of 1956, were historically rooted in decades of struggles over modalities, extents, and severity. The Alkali Inspectorate had controlled SO_2 emissions since the late nineteenth century. Britain hardly ever considered other forms of pollution control. In 1978, an unusual attempt to introduce air quality standards by the Cheshire County Council was "struck down" by the secretary of state in no time (Haigh 1990a, 192). Britain naturally "opposed the establishment of uniform ambient air quality standards" (Vogel 1986, 77).

Early reactions by Parliament to the prospect of having to transpose the directive were systematically negative and framed in terms of the gap between current law and the directive. During a Commons' debate on 18 May 1977, Denis Howell stated:

> The use of air quality standards to be achieved by a stated time represents a new *departure* for us. (as quoted in Haigh 1990a, 185; italics added)

The report by the House of Lords' Scrutiny Committee included in the Annex of the 1976 Fifth Report of the Royal Commission on Environmental Pollution, produced six months earlier, reminded its readers of the successes of the current system and presented reasons for why "we are . . . opposed to the imposition of air quality standards."[64] The actual Report openly discussed how "ambient standards" were "impracticable" and referred to the "difficulties" inherent to "establishing 'bands' of desirable air quality."[65] When the time for transposition came, the 'departure' was naturally resisted by the government.

First, the use of an administrative circular to local authorities was used to safeguard the legal tradition. There would not be a new legal entity that undermined the Alkali and Clean Air Acts. Secondly, the circular would only request, without imposing sanctions, local authorities "to take action thought likely to be necessary." This ensured that very few new objectives would interfere with or overthrow established principles.

Thirdly, the government used the products of its administrative and legal tradition[66] to persuade the European Commission that, indeed, the limit values would be respected. This would rely on current practices to satisfy opposition and otherwise dodge changes in the legal framework. The trick had been foreshadowed in 1976, when the Department of Environment wrote to Parliament that no new legislation would be needed to implement the directive.[67]

Administratively, the *application* of the directive imposed on a rather successful and well-established administrative machinery changes in structure (centralization) and content of responsibility (locally, air quality became the object of control). It also asked for expansion: the CEGB's activities would now have been controlled by the central government. The government opted not to centralize the bureaucracy: circular 11/81 granted responsibility and control to local authorities. The government only partially altered the content of responsibilities: local authorities were charged, without fear of sanctions, to do their best possible to observe the requirements of the directive. The government failed to ensure the submission of improvement plans. The duty of local authorities practically amounted to demonstrating that, thanks to emission controls, air quality was within the acceptable levels set by the directive.

Certain aspects of application were nevertheless easy to realize. Previous efforts at controlling emissions had ensured very good air quality at ground level.[68] The existing policy had produced results similar to those sought by the directive. Thus, in areas where very few changes in the administration would ensure compliance, the government sought to satisfy the European Commission. Monitoring stations were kept with circular 11/81 and used to produce the required reports.

It would take years before the directive would be properly transposed and applied despite the fact that, in practice, its most important requirements had already been attained. The British government rejected at first any proposal related to fixed standards. It was, as a matter of fact, opposed to standards for air quality in general. When Sweden and Norway came to the 1982 Conference in Stockholm armed with data from decades of research showing the British origin of the acidity in their rain, the government and the CEGB continued to oppose proposals of standards for high altitudes. The CEGB, largely responsible for the rain with its high-altitude emissions that brought SO_2 to Scandinavia on eastward winds, replied that the 0.5 g/S/m^3 target would imply "an impossible" 80 percent cut in emissions and would require an investment of over 4,000 million pounds.[69] The Department of Environment, although interested, was overwhelmed by the CEGB and the support given to it by the CBI, at the time afraid of increases in costs of energy.[70]

Once threatened by the European Commission, and aware that a legal and administrative leap could be made (the air quality standards were already in place for much of the country), the British government accepted compliance and passed three successive acts. These acts fully transposed the contents of the directive and promised a proper application.

In sum, the British approach to smoke and SO_2 pollution of air prior to the directive determined the time and extent of transposition and application of the directive. The directive asked for a reorientation of a policy that had focused on emissions for both smoke and SO_2. The reorientation was not impossible: the previous policies had ensured anyway very good air-quality standards. After some years and threats from the commission, the directive was finally transposed and faithfully applied.

7

The Spanish Case

A Favorable, but Statist, Policy Legacy

Spain drafted a faithful translation of the directive one year prior to the country's entrance into the Community with Royal Decree 1613 of 1 August 1985. The decree contained the correct limit values for SO_2 and associated values for smoke, compliance plans for zones in breach of the established limits, a principle comparable to that of the standstill principle, provisions for monitoring stations, and guide values. Transposition was therefore on time and full. Application was, however, partial. Some areas complied; some had no stations and could not at all comply with any of the requirements; a number of areas, with stations, did not comply with most requirements. For those areas in breach of the limits, authorities found it difficult to improve the situation and enforce the decree.

This chapter explains Spain's good transposition and partial application record in terms of the preexisting legal and administrative state policy on air pollution. Transposition was possible because earlier legislation, namely the Air Protection Act of 1972 and Decree 833 of 1975, contained most of the directive's principles, including the imposition of limit values for concentration of SO_2 and smoke and the creation of a monitoring network. Application of the directive was incomplete because a weak administrative apparatus was asked to achieve far more than it could, although the objectives were familiar ones. The administration was weak in two senses: structurally and in its penetration of society. The directive unrealistically asked for the immediate restructuring of the existing administrative layout. Not until the early 1990s would Spain demonstrate some commitment toward administrative reform.

As in Great Britain and Italy, the directive's demand on the organization of interest groups engendered dynamics that were only secondary for implementation. Employers were happy to rely on the state's reaction to the directive to protect their interests. Conservationist groups were almost non-existent and could not mobilize to take advantage of a favorable law.

Time and Extent of Transposition and Application

Royal Decree 1613 of 1985 was enacted one year prior to Spain's admission into the Community. Upon joining the Community, Spain had therefore a punctual transposition of the directive. The decree, as part of its title, stated its purpose: "to establish new standards of quality for air, specifically with reference to sulphur dioxide and particulates."[1] In the third paragraph, specific reference was made to the intent to comply with EC legislation, described as more stringent than current measures. In the fifth paragraph, the decree mentioned the particular legislation in question: "directive 80/779/EEC, on the levels of concentration." The nine articles, the two concluding 'dispositions', and the four tables in the Annex, carefully translated into Spanish law all of the major principles of the directive.

Article 2, Section 1, restated the basic requirement of the directive: "limit values for sulphur dioxide and suspended particulates with reference to the periods and conditions specified in the Annex." Section 2 noted that the value for one pollutant is set in relation to that of the second. Table A in the Annex reported the actual values as shown in table 7.1. With table A, the decree specified the correct relational limit values and absolute limits for SO_2, but not the absolute maximum limits for smoke (the table merely stated that when smoke exceeds 40, 60, or 150 mg/cubic meter then there were associated limits of SO_2). Table B of the decree identified those absolute limits and is reproduced in table 7.2. Together, these tables faithfully transposed the directive's most important requirements.

Articles 5 and 7 laid out plans for those regions exceeding the limits. Those regions would be declared 'Polluted Areas'. Mayors and city governments had to, after a zone was declared polluted by national, regional or local authorities,[2] produce a plan of measures for progressive improvement that included a description of the nature, origin, and evolution of the pollution, and details on the technical strategies chosen to bring levels within the fixed limits. The plan would reach one or more higher authorities, including regional bodies, the Inter-Ministerial Commission on the Environment (CIMA),[3] and the Council of Ministers, for approval.[4]

Article 8 effectively translated the standstill principle found in the directive. It enabled the government to set lower limits than those provided by the directive and the decree. Reasons for setting lower limits included fear of sudden, big increases (as in industrial areas or growing cities) and a desire to approach the guide values.

Table 7.1. Reproduction of Table A in the Annex of Spain's Royal Decree 1613 of 1 August 1985

Period	*Limit Value for SO₂ (mg/cubic meter)*	*Associated Limit Value for Smoke (mg/cubic meter)*
Annual	80	> 40
	120	≤40
	[means of daily averages registered during the year]	
1 October to 31 March	130	>60
	180	≤60
	[means of the daily averages registered during the six-month period]	
Annual (composed of 24 hr. periods)	250—not to be exceeded for more than 3 consecutive days	>150
	350—not to be exceeded for more than 3 consecutive days	≤150
	[98th percentile for all the average daily values registered during the year]	

Article 3 identified guide values for the future. "By guide values," the passage noted, "is understood concentrations of sulphur dioxide and suspended particulates in reference to periods and conditions as outlined in Table C and D of the Annex." Table 7.3 reproduces tables C and D of the decree. These guide values fully complied with those set out in Annex II of the directive.

The meticulous transposition of the directive, unmatched in Great Britain or Italy, was followed by a less impressive applicative phase. The most basic tool needed for application was a proper national monitoring system, capable of supplying the basic information that would identify the status of pollution throughout the territory and permit the formulation of policies that would satisfy the remaining requirements: claim standstill zones, identify polluted areas, formulate improvement plans, and work toward guide values. Article 9, Section 1, indicated that an existing network, dating from the Air Protection Act of 1972 and applicative decree,[5] would supply the essential information:

Table 7.2. Reproduction of Table B in the Annex of Spain's Royal Decree 1613 of 1 August 1985

Period	Absolute Limit Value for Suspended Particulates
Annual	80 [mean of the daily averages registered during the year]
1 October to 31 March period]	130 [Mean of the daily averages registered during the six-month]
Annual (composed of 24 hr. periods)	250 [not to be exceeded for more than 3 consecutive days] [98th percentile of the daily averages registered during the year]

Monitoring atmospheric quality, the promotion and coordination of which is the responsibility of CIMA, will take place starting from the data that the National Network for the Monitoring and Forecasting of Atmospheric Pollution will provide.[6]

Acknowledging the backwardness of the network, especially the small number of stations and the large percentage of manual stations,[7] legislators urged for improvements in Section 2 of Article 9, requiring the construction of "additional stations near areas rich in polluting sources." The second and final disposition specified those responsible for the upgrading:

The Ministries of Public Works and Urbanism, of Industry and Energy, and of Health will be given instructions to homogenize, coordinate and optimize with national character the equipment and the use of the stations . . . the Ministry of Economy and Treasury will provide the funds necessary for the execution of this Royal Decree.

In practice, the improvement of the network, despite these recommendations, barely occurred. The government itself recognized that the replacement of manual stations had yet to occur as of 1990, due to unexpected problems and lack of funding.[8] For the same reasons, the network hardly expanded into any new area.

Equally problematic was the fact that in those areas where data was available and limits were exceeded, cooperation from industrialists and other polluters was truly minimal. Systematically, polluters ignored the decree. With equal consistency, the government failed to punish those that would not comply. Despite enforcement and sanctions specified by the 1975 decree and

Table 7.3. Reproduction of Tables C and D in the Annex of Spain's Royal Decree 1613 of 1 August 1985

Period	Guide Value for SO₂ (mg/cubic meter)	Guide Value for Smoke (mg/cubic meter)
Annual	40–60	40–60
	[means of daily averages registered during the year]	
24 Hours	100–150	100–150
	[daily averages]	

considered valid by the 1985 decree,[9] only in 1990 was the first violator actually tried and found guilty.[10] As a result, a number of contaminated areas remained as such for years after being identified.[11] To a large extent, the principles of the directive were not applied.

Why did Spain have such a precise transposition but partial application? The next section investigates this puzzling performance.

Legal Antecedents:
The 1972 Air Protection Act and the 1975 Royal Decree

The proper transposition of the directive in Spain entailed a minor conceptual transformation of two preexisting legal entities, the Air Protection Act of 1972 and the Royal Decree of 1975, and was for that reason possible. Unlike the legislation of Great Britain prior to the directive, the Air Protection Act of 1972 and the Royal Decree of 1975 focused on both emissions *and* concentration levels. They prioritized, moreover, specifically concentrations of SO₂ and smoke. The difference between the directive and the existing act and decree was in terms of levels: the directive imposed stricter levels. Because the new levels did not threaten the legal legacy of the country, Spain could easily draft a new decree openly designed for the directive. A proper application of the directive, on the other hand, required a much more capable administrative system than the one Spain had at the time; in addition, it asked for a radical transformation of the relationship between a statist government and interest groups. The Spanish government could only begin to move toward proper application in the early 1990s. This section examines in detail the Air Protection Act of 1972 and the decree of 1975, the legislations that preceded the 1985 decree.

The 1972 act was the first major legal entity concerned with controlling air pollution at the national level.[12] It was designed by the Comisión Técnica Asesora Sobra Problemas de Contaminación Atmosférica, instituted in 1969 and dissolved in 1975 (Mateo 1992, 297n10).[13] The act interpreted the problem from two angles. Similar to the British approach, it understood emissions to be a central feature of pollution: pollution was defined in part as emissions of harmful substances into the air. More fundamentally, pollution was also seen as a problem of the concentration of harmful substances in the air. Several references to the 'saturation' of the atmosphere as *the* problem at hand characterize the act's opening paragraphs. The beginning of the fifth paragraph is in this sense typical:

> The saturation of the atmosphere—that is, the exhaustion of all possibilities of assimilation of new polluters in light of the maximum standards legally permissible—produced by those emissions of polluters from activities in a certain area, must live up, to justify its existence, to an improvement in the lifestyle of the majority of people living there.

Because emissions and air quality were inevitably linked, the law proceeded to define "the struggle against atmospheric pollution" as having "two essential aspects:" that of the "the defense . . . of the quality of air through the imposition of adequate concentration standards," and that of the establishment of maximum limits of emissions for polluters, designed especially for power stations and industrialists.[14] The "solution to the problem" therefore began, naturally, with the government's ability to promulgate "general guidelines for setting air quality standards, [*and*] levels of emissions of polluting substances, quality of the fuels . . . and controls for heat generators." What these levels would actually be was a matter taken up by a later decree, as requested in the first disposition.[15]

In a similar fashion to the directive's demands, the act specified fairly detailed plans of compliance for those areas that would not meet the prescribed standards. Article 6, Section 1, laid out the general principle: "those areas declared polluted will enter a special regime of action aimed at the progressive reduction of concentration levels until actual levels reach those prescribed by the law." Sections 2 and 3 outlined how the principle would be respected. There were two possible venues for bringing pollution down to acceptable limits.

First, the government, after informing provincial commissions and having consulted the affected town councils, was authorized to impose all or some of the following measures:[16] (*a*) use of less polluting fuels by power plants and communication to the authorities of what these fuels might be,[17] (*b*) prohibiting the installatation of new incinerators that cannot comply with emission standards, (*c*) the obligation that new heating plants installed during this critical period use nonpolluting sources of energy, (*d*) the enforcement of all necessary measures that could ensure the reduction or purification of pollutants exiting from chimneys. A second, more drastic, option could be pursued: it entailed the prohibition of new installations and the dismantling of existing ones. In most cases, as stated in Article 11, funds would be provided to offset parts of the costs that polluters faced when trying to observe the law.

The act also contained principles very similar to the guide values and standstill principles found in the directive. Article 3, Section 2, specified that

> the government will be able to establish limits of emissions *more strict* than the general ones when, after observing and analyzing the circumstances, it will be decided that persons or goods in the area are threatened. (Italics added)

In its conclusion, the 1972 act provided for a national monitoring network composed of mobile and fixed stations. Article 10 decreed that

> the government will establish a national network of fixed and mobile stations for the monitoring and forecasting of atmospheric pollution; the network will be administered by the Ministry of the Interior.

The idea of creating such a network was not novel, since a number of cities, including Barcelona, Madrid, and Bilbao, had already built their own for the explicit purpose of measuring levels of SO_2 and smoke.[18]

These were the formidable (for the time and in light of the rest of Europe) propositions of the act. With its two-pronged approach, and its provision for a national network, the act was certainly one of the most comprehensive legal approaches to pollution in Europe. With a two-year delay, the government proceeded to fill in with specific quantities and measures the general framework thus created.

Royal Decree 833 of 6 February 1975 was designed for the sole purpose of realizing the 1972 act.[19] With its title "For the Realization of Law 38 of 1972 (22 December) for the Protection of the Environment," the decree methodically substantiated the claims of the act. Perhaps most importantly for the future implementation of the directive, Article 4, Section 1, approached the issue of setting the actual limit values:

> In conformity with Article 2 of Law 38/1972 of 22 December, on the Protection of the Atmosphere, the levels of concentration . . . for what is admissible, as well as for what is to be considered a polluted zone . . . are detailed in Annex 1 of this Decree.

Annex 1 specified, first, levels of SO_2 and, secondly, levels of suspended particulates. Other pollutants followed. In all cases, and in a manner similar to that of the directive, instructions were given to calculate different types of concentration. Formulas for calculating acceptable median concentrations

Table 7.4. Concentration Limits for SO$_2$ and Suspended Particulates in Spain's Decree 833/76

Period	SO$_2$ (mg/cubic meter)	Suspended Particulates (mg/cubic meter)
One Day	700	300
One Month	256	202
One Year	150	130

were laid out for the following periods: two hours, one day, one month, and one year. The concentration levels for admissible and polluted areas for SO$_2$ and suspended particulates were specified in Sections 2 and 3 of Annex 1. They were less strict than those of the directive by some degrees and they were not set in relation to each other, but the categorization of the different types of concentration, for one day and for the year, was very similar to that of the directive. Table 7.4 shows the actual quantities. Articles 14 through 32 specified additional measures for what should be done in regions where limit values were exceeded. Articles 16 through 21 clarified who could declare a zone polluted. Any person, private or public, could bring to the attention of local or regional authorities the high level of pollution in an area. Mayors were responsible for drafting proposals for action; their proposals would then reach the following organs: governor of the province, Ministry of the Interior, General Directorate of Health, Inter-Ministerial Commission on the Environment, and Council of Ministers.

Articles 22 through 24 specified what actions would be taken as a result of the declaration. All of the steps, previously analyzed, of Article 6, Sections 2 and 3, of the 1972 act would apply. Partial funding would also be distributed as stated in Article 11 of the 1972 act. There was in addition a number of new measures. Areas affected would be entitled to have monitoring stations from the national network. Mayors were required to start Centers Against Atmospheric Pollution dedicated to study, inform, and propose to the government plans for improvement. Industrial plants would receive instructions on what fuels they were allowed to use. Power stations with production exceeding 2000 therms per hour were required to use clean fuels.[20] Industrialists could be asked to alter the height of chimneys. Finally, the government would decide when and if to declare an area free of pollution, following a set consultation pattern with the Inter-Ministerial Commission, local authorities and trade representatives.

Articles 6 through 13 contained provisions for the national network. The network would be managed by the Ministry of the Interior and would be used first of all to collect data on SO$_2$ and suspended particulates.[27] It would have a national center, sitting in the General Directorate of Health, regional centers for the collection of data, and regional centers for the analysis of data. 'Regions' in this case referred to natural, geographic regions, not necessarily overlapping with existing political regions.

The national center's duties were outlined in Article 8. Its major responsibility was one of coordination and general policy-making: to request, from the centers for the analysis of data, and to review reports on the state of atmospheric quality, and, upon reviewing the data, to present to the Inter-Ministerial Commission and relevant bodies a general picture of the situation. The national center was expected to indicate in its reports which areas were at risk and to propose how the national network should be expanded.

The duties of the regional centers for the collection of data were outlined in Article 9. These centers were expected to gather data and "forward it with the most expedient means" to the appropriate authorities. The duties of the regional centers for the analysis of data were outlined in Article 11. Article 11 specified the modalities for communicating the findings.[22]

Decree 833 dedicated much attention to emission controls as well. It specified limits, criteria, types of polluters, and administrative actions for ensuring that emissions would be kept under control. The relevant articles will not be examined here, since they bear little weight on the role of the decree for the implementation of the directive. The decree did not elaborate any further on Article 3, Section 2, of the 1972 act. The original dispositions remained therefore unaltered.

The Air Protection Act of 1972 and the 1975 decree were the Spanish answer to the problems of pollution: they were the choice of Spanish legislators in one policy arena. Drafted many years before Spain joined the Community, they followed domestic initiatives, and were not the result of pressures from the international arena. Rooted as they were in the Spanish legal mentality, they would later play an important role for the transposition of a foreign law, the Air Pollution Directive, that carried with it remarkably familiar demands.

Administrative Antecedents: The Structural Weakness

Between 1975 and 1985, the Spanish government could not translate into practice many of the ideas of the 1970s legislations. As Mateo stated:

> Our legislation of 1972–75 began with the double strategy of controlling emissions and concentrations, but was followed by poor application. (1992, 299–300)

There were two general and probably related reasons for the disappointing performance: the weak structure of the administrative apparatus, examined in this section, and the poor penetration by the state of society, examined in the next section.

Critics and governmental reports alike acknowledge the structural administrative limitations that accompanied the application of the 1972 act and 1975 decree. The administration had failed to build a sufficiently extended network, to create a Ministry of Environment or a centralized unit, and to enforce the established standards in those regions documented as highly polluted.

The Red Nacional (the National Network) boasted a small number of technologically limited stations located around some cities. For financial and operational reasons, most of the existing stations were manual,[23] at a time when Germany and Great Britain had extensive automatic networks.[24] As such, manual stations could collect small amounts of data in a time period and required long periods and resources for the analysis of data.[25] Annual averages could not be easily collected, and were generally approximated from measurements of shorter periods.[26] The government itself admitted the limits of manual stations and noted, in 1977, that provisions would be made to equip those areas documented as polluted with automatic machines (Presidencia del Gobierno 1977, 785). Replacement, however, barely occurred.[27] By 1985, several industrial regions were without a station, while most existing stations could only collect insufficient amounts of data.[28]

Surprisingly for a country with as serious and complex a legal measure against pollution as the 1972 act (and its implementing 1975 decree), Spain had no major administrative organ purely designed for environmental purposes. The 1960s witnessed the creation of sectoral bodies, such as the National Commission for the Prevention of Contaminated Waters (in 1962) and the Technical Commission of Industrial Pollution of the Atmosphere (in 1969). The 1970s witnessed the creation of more general bodies that were always short, however, of extended powers. In 1972, the CIMA and the (never operational) Commission for the Environment to the Government were created to carry out the 1972 act. In 1979, in light of the failures of CIMA, a Directorate General for the Environment was established within the Ministry of Public Works and Urbanism. Finally, beginning in 1979, steps were taken to create a Commission on the Environment within Parliament.[29]

In the midst of this confusion, responsibilities for carrying out the 1972 act fell in the hands of different ministries. The act's attempt to tackle both air quality and emission standards surely made matters worse.[30] Functions, at any rate, were never clearly delegated. Silvers' observations were made in 1991 but described a situation that had not changed since the 1970s:

> Because so many different Spanish ministers are responsible for different aspects of environmental protection, an overlapping of

authority exists. Contributing to [this problem] is Spain's recent failure to elevate the environment to ministry status. (Silvers 1991, 308)

In 1979, some functions were eventually centralized in the hands of the Directorate General for the Environment. The directorate had, however, a very limited budget, staff, and legitimacy; its capacity to function effectively was therefore heavily curtailed (Silvers 1991, 308).

The lack of a powerful administrative center precluded the possibility of coordinating and funding the expansion of the network, and monitoring and enforcement activities. By 1979, it became in fact obvious that, in those areas where data on excessive pollution were available, the enforcement mechanisms were almost nonexistent. The 1977 report by the government listed thirteen cities in which acceptable limits for SO_2 and/or suspended particulates had been exceeded (Presidencia del Gobierno 1977, 799–901). Only in 1990 was the first offender convicted.[31] "In cases of industrial pollution," stated a report, "the government has appeared reluctant to prosecute" (Luke 1992, 12). The absence of any major conviction against polluters could itself be interpreted as an indication that compliance, following perhaps a British model of bargaining and informal agreements, followed. Yet evidence suggested otherwise: the 1977 report noted how little industrialists had opted to respect the new measures, and how they "had adopted scant measures to limit their emissions of polluting substances" (Presidencia del Gobierno 1977, 210).

A list of related explanations could account for the poor monitoring network, the absence of a Ministry for the Environment, and the hesitance to enforce the 1972 act. The transition from a statist to a democratic government, some have noted, was possible only with the partial appeasement of conservative elements in society, which included most industrialists; thus, in the late 1970s and following Franco's death, transition governments were still inclined to please right-wingers and could not adopt an environmental policy (Luke 1992, 12). Violators of the law were not persecuted, and the administration received little guidance or support to carry out its functions.

Others, such as Secretary of State for the Environment Albero in 1991, have directly pointed to the lack of financial and conceptual commitment toward environmental policy, itself a function of Spain's preference for economic development over the environment.[32] These were the years of the Spanish 'economic miracle', the long awaited and unprecedented economic growth that no one would think of obstructing for hitherto irrelevant issues like the environment.

The weakness of the act itself also played a significant role. As Silvers noted, on paper the act was a rather impressive law, with serious intentions to impose limits on pollution; in reality, however, "the Act was riddled with loopholes and it was soon taken advantage of by eager and hungry capitalists" (Silvers 1991, 290).[33] This could explain, among other outcomes, the courts' hesitance to enforce the law.

Regardless of its sources, the structural weakness of the administration would prove decisive for the application of the Community directive. Its effects are examined later in this chapter. The next section considers a second, nonstructural weakness of the administration: the relationship between the state and affected interest groups.

Administrative Antecedents: The Exclusion of Social Actors

The impetus to undertake a serious environmental policy in Spain may have originated from events in the social sphere, such as the death, due to air pollution, of three people in 1968 in Erandio and some protests by the population against industrialists and their polluting activities.[34] Yet the Air Protection Act of 1972 and most environmental policies of the 1970s did not actually originate from forces emanating from civil society. They were primarily state-induced initiatives that excluded, both in the drafting and administrative aspects, relevant interest groups. Once excluded from the policy-making process, groups were also left out from the administrative phases. Industrialists thus found the laws too demanding and too sudden and unrealistic, and therefore opted to ignore their requirements.

"Spanish environmental policy," wrote a major environmental researcher, "has been to all intents and purposes a task of the government." She continued: "in its formulation no institutional cooperation with industry has been set up, nor have groups of experts been called upon," (Aguilar 1993, 236) and, "the environmental political process of Spain [has gone on] without taking private actors into account" (Aguilar 1993, 225). The 1972 act was not, in fact, proceeded by large scale studies designed to predict the impact of reducing industrial pollution on the economy and to consult, beforehand, with trade representatives.[35] The 1975 decree did not include, among the list of practical measures designed for implementation, any contribution from experts or representatives.

Nor did these legislations announce the formation of new consultative organisms that would include private representatives. A handful of projects were forwarded in the late 1970s and early 1980s for the creation of such organisms but were disregarded systematically.[36] A minor body, the Committee for Public Participation, came into existence in 1983 but was eventually suppressed three years later.[37]

Naturally, industrialists and directors of power plants found it difficult to comply with the laws and listed, among the most serious obstacles, issues that would not have been a problem in a more cooperative context. The exclusionary practices of the state had not penetrated society, as in Great Britain, and had not led to those valuable formal or informal procedures between officials and affected individuals and companies that could ensure, before the law was even enacted, compliance.[38] Not surprisingly, industrialists justified thus their failure to observe the law on the following premises: the law was excessively rigid, compliance required alternative uses of energy, but no personnel was available to provide the knowhow, and compliance required technical knowledge of emissions but no personnel was available to supply the knowhow.[39] These problems could have been resolved if prior consultations had been set up between government and affected parties.[40]

There were of course reasons for the state's exclusionary approach to environmental policy-making. Franco's government in most policy areas acted without representation of interest groups or members of society. As Linz put it in the midst of a comprehensive discussion about Franco's regime and its relationship to civil society: "in the world of interest representation, as in the political system, there was limited pluralism, one that excluded any autonomous organization of the working class and limited even conservative forces . . . interest politics . . . was subordinated to Franco's ultimate power" (1981, 391). In the case of environmental policy, interest groups were especially left out: the 1972 act and 1975 decree represented the responses of a government that had suddenly realized the incredible polluting side-effects of great economic growth, and had decided, resolutely, to act immediately, and then later realized that costs of application would be too high for development and the transition to democracy.

Explaining Implementation

The policy legacy of the Spanish state accounts for both the positive and negative aspects of the implementation of directive 80/779. Transposition of the directive posed no challenge to the legal principles of Spain: for this reason it occurred faithfully and on time. The directive introduced stricter limits, but espoused the same conceptions of pollution control: ones that focused on air quality. The proof that legislators themselves understood the directive as conceptually akin to existing legislation is found in the opening paragraphs of the 1985 decree used to transpose the directive and entitled, in part, "[A Law] that Establishes New Rules for the Quality of Air in Terms of SO_2 and Suspended Particulates." The first paragraph opens with the recognition of existing norms:

> Law 38/1972 of 22 December on the Protection of the Environment established in Article 2 that the Government will set levels of concentration, meaning with such terms the maximum tolerable

present in the atmosphere of each pollutant . . . Decree 833/1975 of 6 February specified in Annex I the levels.

The second paragraph starts by noting that "various circumstances justify a *partial modification* of Decree 833/1975" (italics added). The third paragraph mentions the overriding 'circumstance': "Community norms." A direct comparison between the quantities laid down by the 1975 decree and those of the directive shows the relative small gap in concentration levels between Spanish and EC law.

A simple qualitative comparison shows that the 1972 act and the 1975 decree had formulated compliance plans for regions in breech of limit values, guide values, and standstill principles. They had also provided for a monitoring national network. Transposition could be faithful and carried out even before Spain joined the Community.[41]

Application required a powerful administrative apparatus and a collaborative state-interest groups relationship (or a state strong enough to impose the new rules). Spain's entry into the EC pushed Spain to adopt the laws and norms of the Community,[42] but it also brought the country to face peculiarities and inefficiencies rooted in the nation's history. The weaknesses of the state administrative apparatus during the 1970s had persisted up to 1985 and had precluded the proper application of the directive. The directive unrealistically asked for the quick removal of these weaknesses.

In 1990, a major international study on the implementation of the directive found that only 30 to 40 percent of stations belonging to the National Network could provide reliable data and that plans for modernization were lagging behind schedule (Alvarez 1991, 62). Data from the Madrid area itself, one of the best equipped, was being questioned by the EC. Suspicion was so high, in fact, that in the early 1990s the EC decided to execute a six month monitoring program to ensure that Madrid's network of twenty-one stations recorded faithfully pollution levels and that levels themselves were below the limits established by the EC (Silvers 1991, 310).[43]

Difficulties with finances and the coordination of central-regional activities were cited as the primary causes of the underdeveloped network:

> The process, according to government statements, is proving to be longer and more complex than expected because the Ministry of Health, which is in charge of the entire operation, wishes to place the stations in the hands of the autonomous communities which lack the funding needed for modernising and maintaining the network . . . the Directorate-General for Environment is currently trying to modernise the network. (Alvarez 1991, 62)

But financial and central-regional problems, rather than being the cause of the problem, were actually the symptoms of a larger problem: the European Community had asked the Spanish administration to do what it, in the 1970s, could not do. The weak monitoring network of the 1970s was the same network charged in the 1980s with new Community expectations: the problems afflicting the network had not disappeared, nor were they about to disappear upon the introduction of the directive. Those problems had come about for historical reasons, in part related to Spain's quick and belated industrialization process and the preference of government officials to favor development over environment, and in part related to the transition from Franco's regime to more democratic governments and the choice by officials to appease, as much as possible, the conservative elements of Spanish society.

Besides monitoring stations a second structural weakness continued to hamper the applicative process of the directive. Despite efforts to centralize control over environmental policy in one single administrative unit, Spain continued to have no Ministry of Environment well into the 1990s, a situation that made the application of the directive practically impossible. Responsibilities continued to be ill-distributed among various units: the choice to shift responsibilities for the National Network from the inapt Ministry of Health to the Directorate-General for the Environment exemplified this. The application of the directive, one of many environmental directives that Spain had to observe upon its entrance into the Community, required without question a central unit. Yet Spain, in 1992, was still

> the only Community nation with no ministry exclusively dedicated to the environment. Albero, Secretary of State for the Environment, holds a rank equivalent to a British junior minister and his department reports to the Ministry of Public Works and Transportation. (Luke 1992, 13)

The reasons for this unique shortcoming in Europe were the same that explained the weak National Network. The consequences affected more than the upgrading of the network. Information dissemination, consultations, and enforcement wherever applicable proved nonexistent.

As it happened for the 1972 act, the application of the Air Pollution Directive witnessed no enforcement. As of 1993, only one polluter in breach of the law had been found guilty (in 1991).[44] The conviction was recognized at the time as a signal that indeed the administrative system, along with the courts, was moving toward a much stricter regime. More time and efforts, however, were still indispensable ingredients for the change to come to completion: the directive asked for remarkable *changes* in the relationship between irreverent polluters and the administration, in the relationship between the administration and the judicial system, and in the judicial tradition itself.[45]

The directive also asked that polluters and administration cooperate at some levels.[46] In the 1990s, however, Spain was only beginning to build institutionalized forms of participation for affected interest groups. A number of regional associations sprouted in the late 1980s. In 1988, for instance, the Basque government and the Basque Business Confederation met for the first time, soon after a cooperation agreement between the Environmental Agency and the Industrial Organisation of Andalusia (Aguilar 1993, 240). Contracts between the government and large firms for the design of environmental adaptation plans also came into existence for the first time in 1992. Nonetheless, as Aguilar herself points out, these were only the first steps. Still absent was a permanent dedicated central structure for routine industrial participation, or some mixed commission of various interest groups for the design of a national environmental program (Aguilar 1993, 237).

The reorientation of a tradition of exclusion to one of cooperation could only take time and resources. The early 1990s saw the beginning of a gradual transformation in the legislative and administrative branches.[47] Yet exclusion still belonged to the statist approach of the Spanish government, especially in the sphere of the environment, where for decades industrialists did not attempt to cooperate for as long as government did not enforce its laws.

In sum, the Spanish approach to smoke and SO_2 pollution prior to the directive determined the time and extent of transposition and application of the directive. The directive was consistent with the legal framework of Spain and was thus transposed fully and on time; it asked for a change in most aspects of the administration, and was thus only partially applied. The change toward a more capable administration would take time and resources: at the beginning of the 1990s, Spain was only taking the first steps toward that transformation.

8

Conclusion

The Future of a United Europe and Theoretical Considerations

The journey here comes to an end. It followed the life of two directives in six countries, exploring, as a way of understanding the turn of events, decades of history, struggles, and resolutions. It is certainly difficult to draw general conclusions from this journey when the European Community and Union have promulgated hundreds of directives. The depth of knowledge required for a study of the factors influencing the implementation of two directives naturally limits the number of case studies that can be considered here. Some lessons can nevertheless be drawn from the evidence presented. They concern the implications of the findings for the future of a European transnational market and of transnational markets in general; and they concern the implications of the findings for political science and sociological theory.

Summary of the Findings

The previous chapters' main purpose was to illustrate, with as much attention to details as possible, that the creation of the European transnational market requires, despite the simple request of its directives, the transformation of deeply rooted institutions and, by reflection, practices in some or most member states. In particular, I turned to the Equal Pay and the Air Pollution directives to offer a clear picture of what institutions are actually being targeted and the degree to which these institutions are rooted in the economic, political, and social history of nation states. The directives, I argued, challenged two domestic institutions: the policy legacies (understood as the legal and administrative entities that exist in a country prior to and at the time that the directive is enacted) of member states and the existing organization of interest-groups (understood as the distribution of goods and resources among groups in society). These institutions are the result of decades of political, economic, and social struggles, interpretations of reality and resolutions. The overview of these directives' fates in six member countries was also intended to prove a second point: that the success or failure of the Union's directives in member states was determined by the extent to which the directives challenged those domestic institutions. Implementation suffered when a directive asked for change in the organization of interest groups or for a change in the legal and administrative tradition of a country. When, on the other hand, a directive brought principles consistent with, or that strengthened, the current organization of interest groups or the legal and administrative tradition, implementation benefitted.

In some cases, interest group organization and policy legacies were affected in a similar way by the directives considered.[1] In Italy and Great Britain, the Air Pollution Directive challenged existing laws and interest group practices. The Equal Pay Directive was not implemented in Great Britain because both policy legacies and the organization of interest groups would have been greatly affected. In France and Italy, however, the Equal Pay Directive had a more contradictory relationship to interest group organization and policy legacies. In France, the directive asked for a small change in the legal tradition of the French state in the area of female labor, but a strong change in its administrative apparatus and the organization of women. In Italy, the directive brought novel legal and administrative principles but entailed little reorganization of working women, who were already benefiting from equality rules in collective agreements. Similarly, the Air Pollution Directive asked little from the legal tradition of the Spanish state, but posed great demands on its administrative apparatus and interest groups. In all of these cases, opposing influences faced each other. Outcomes generally depended on the relative strength of these influences. The organization of Italian women was already so close to the idea of the Equal Pay Directive that a legal change, although a departure, could anyway be introduced; building an administrative structure to oversee the new law was too much of a departure. In France, the state had already taken away from unions and management control over female labor. The strong presence of a legal and administrative tradition in the field of female labor ensured the overcoming of any opposition unions or management could foster. In Spain, opposing forces influenced implementation of the Air Polution Directive at different levels: transposition could take place but application would not follow.

Table 8.1 summarizes how the Equal Pay Directive affected national institutions in each country. It then indicates how the directive was implemented. Admittedly, table 8.1 encompasses too many dynamics between the directive and national institutions to be self-explanatory. A brief recapitulation of the main findings is therefore in order.

Table 8.1. Implementation Records in Light of Demands of the Equal Pay Directive on National Institutions

	Equal Pay Directive		
	Did it asked for a shift in current policy legacy?	Did it ask for the reorganization of interest groups?	Was it implemented?
France	Administrative shift	Yes	Transposed / applied late but fully
Italy	Legal and administrative shift	No	Transposed / not applied
Great Britain	Controversial legal and administrative shift	Yes	Not transposed / not applied

In France, there existed a strong legal and administrative legacy in the field of female labor resulting from unique circumstances (the feminization of the workforce, the decision of the state to intervene, and the primarily conservative nature of that intervention). The state had already produced an advanced law on equal pay and eventually came into possession of the administrative capacity needed to apply such a principle, after some conservative elements had precluded proper application. The directive could thus be easily transposed and, with some delays, applied. The directive did upset interest group organization: it did, after all, award wage equality to underpaid women. Management, however, proved incapable of opposing the positive influence of the state, while organized women, as predicted, could do little to capitalize earlier on the presence of the directive.

In Italy, women had already won, in collective agreements, equality of pay. The demands of the EPD would have had little impact on Italian society. Although legally a challenge, the EPD's overall irrelevance for Italy induced the state, which had no antagonist legal measure, to comply with transposition. For transposition, then, the presence of collective agreements obtained through strong organization influenced outcome more than the lack of state laws on that matter. The state did not comply when implementation implied significant changes for itself and society: it refused, and proved incapable, of creating the administrative mechanisms that would transfer the authority over female and male wage differentials from trade unions and business organizations to the state.

In Great Britain, the directive would have brought great changes in the organization of women and men, and in the legal and administrative tradition

Table 8.2. Implementation Records in Light of Demands of the Air Pollution Directive on National Institutions

	Air Pollution Directive		
	Did it asked for a shift in current policy legacy?	*Did it ask for the reorganization of interest groups?*	*Was it implemented?*
Italy	Legal and administrative shift	Yes	Transposed very late/ not applied
Great Britain	Controversial legal and partly controversial administ. shift	Yes (although minor)	Transp. very late/applied partly on time, fully very late
Spain	Only administrative shift	Yes	Transposed/very poorly applied

of the country. In 1975, the state would not alter its legal and administrative structures (which were essentially antagonistic to equality of pay and products of under-representation), and organized women could not mobilize to ensure transposition and application.

Table 8.2 summarizes the demands of the Air Pollution Directive on national institutions and relates outcomes to those demands. For this directive, the policy legacy of each country proved especially influential on implementation. The table illustrates very clearly the impact of policy legacies on implementation, even if, at least in theory, interest groups might have intervened. In all cases, the directive's demands on the policy legacy of a nation proved crucial for outcomes.

Italy's fragmented legal and administrative apparati, primarily focused on emissions, would make the transposition and the application of the directive an impossible demand on the state. The absence of interest groups, such as green organizations, responsive to environmental issues explained in part the poverty of the legal and administrative apparati, and the absence of any pressure on the state in 1980.

In Great Britain, a long tradition of controlling emissions, rather than air quality, proved fatal for the timely transposition of the directive. An equally old tradition of decentralized and informal administrative control over pollution proved in part detrimental to the creation of an administrative entity capable of imposing fixed standards nationwide. The administrative structure did however ensure that the principles of the directive be respected at least in some areas on time. Good air quality and an extensive network eventually led to full, although belated, compliance.

Spain's early law on air quality standards made transposition possible. The applicative shortcomings of that law, embedded in state-society relations and the timing of industrialization in Spain, explain the poor application of the SSD.

Implications for the Creation of Transnational Markets and the Future of the European Union

Is a united Europe possible? What stands in the way of unification? What lessons can be learned from the European effort about the process of creating transnational markets? The study of the implementation of European directives suggests that, for a united Europe to become reality and for a transnational market to be created, the history of individual nations must, in a sense, be overcome. The EC has sought, and the EU will seek, to homogenize various aspects of European societies, beginning with legal and administrative contexts and ending with political, social, and economic realities. Institutions, as defined in this volume, have been the primary target. They have resisted directives carrying significant change, pointing to their nature as very stable social entities, reflecting decades of struggles between social forces and normative ideas about the organization of the economy, the polity, and society. Naturally, the question of whether Europe will ever be united, or of how change can take place, arises. The model offered here seems to depict a static scenario: directives that introduce change will be rejected by nations. This is by no means a conclusive finding. In at least three ways, the evidence presented in this volume points to the possibility of change.

First, in Great Britain, Italy, and Spain, the Air Pollution Directive did challenge national institutions but was at least partially implemented. Great Britain and Italy eventually passed proper legislation, even if doing so posed a challenge to them and their interest groups. Spain pushed to improve its administrative system, even if in the past it had failed to do so. At least in some cases, then, directives that posed a challenge to national institutions were not implemented fully and on time *but* they were, nevertheless, eventually partly or fully implemented.[2] Directives in these cases provided an incentive for mobilization, either on the part of interest groups or the state. With time, and following, surely, some internal changes, these directives underwent partial or full implementation, as groups and relevant state actors learned about the new legislation and mobilized to seek its implementation.

The evidence suggests that the EU can, in addition, take steps that will improve the implementation of future directives. The European Commission could design directives in light of the institutional context of each member state. Drafting should follow closer interaction with representatives of affected interest groups, legislators, and administrators of different nations. At present, much attention is paid to representatives of interest groups and little attention is given to the gap that reigns between plans for a directive and national legal and administrative institutions. At least four of the six case studies considered in the preceding chapters pointed to the importance of national policy legacies.

The European Union, moreover, could dispose of more severe enforcement mechanisms. At present, in at least two ways the Union lacks serious measures. First, penalties for noncompliance are not sufficiently high. Often, in the case of great gaps between the directive and domestic institutions, noncompliance is less costly than compliance. Building a new administrative structure in Spain for ensuring the implementation of the Air Pollution Directive and imposing the new principles to industry would have costed the country enormous sums. Penalties from Brussels would come late and would be, if actually paid, minor in comparison.

Secondly, the European Court of Justice (ECJ) could have a more clearly defined mission and jurisdiction. There is much confusion about the relationship between the ECJ and national courts. What should the ECJ do when national courts fail to prosecute? Should the ECJ's interpretation of directives be more legitimate than that of national courts? What cases are allowed to reach the ECJ? There are no answers to these questions, only great scholarly and legal debates. A clear mission and jurisdiction are however necessary to ensure homogeneous, predictable enforcement of directives.

This volume leaves open the normative question of whether unification is good or bad. It does note, on the bases of the findings, that unification implies or actually requires the partial homogenization of European countries. By reaching deeply in the legal systems of member states, directives set the common conceptual grid upon which the economies, polities, and societies of hitherto very different peoples will be built. In a sense, unification requires the rejection of nations' histories. This, European legislators and citizens alike, must become at once conscious of and willing to accept.

Alternative Explanations and Future Research

Admittedly, this volume has offered only one instance of transnational markets and has focused on only one (the implementation of directives) of the many ways in which those markets seek entry into nation-states. It could have, for instance, approached the problem from a more traditional economic point of view and have argued that transnational markets reach nation states via the internationalization of capital, and that understanding the forces driving this process would do much to improve our understanding of such markets.

I find it more interesting, however, to consider alternative explanations for the phenomenon I chose (the implementation of laws) as a way of challenging my argument in the preceding chapters. Indeed, however compelling, the curious reader would be right to wonder whether in fact this is the only explanation for the implementation of EU law available. As it turns out, literature on the EU is devoted primarily to analyzing various aspects of *policymaking*. The Council of Ministers, the Commission, the European Parliament, the Court, various committees, parties, interest groups, voting procedures, and charters are studied for their role in the formulation, discussion, and production of policies.[3] Factors outside the formal institutional boundaries of the Union influencing policy-making are also studied: the impact of national political parties, national labor unions, national economic policies, national class pressures, and regional disparities have received considerable attention.[4] Finally, the autonomy of the European Union, strategies for integration, supranational and federal features, and possible future directions have been the subject of debates for many years.[5] In contrast, as discussed in the first chapter, only a few works have discussed to some length the problem of implementation. Some are informative but do little to explain the causes of failed implementation.[6]

From the remaining works, however, it is possible to construct what might be alternative, competing explanations for implementation. These explanations focus on the role of EU-level variables (improved formulation of directives, fluctuating international recognition of the EU) and domestic variables (public identity, the role of leaders, and the efficiency of the state machinery). With the exception of two, all can be fairly safely dismissed.

Over time, for instance, the Union has made an effort to design directives that imply fewer and more precisely defined changes for industries and all affected parties. For this purpose, jurisdiction over the formulation phase has been given to bureaucrats from national administrations competent in the specific areas covered by the new law.[7] On behalf of the EU, these administrators consult regularly with industries and interest groups to draft laws that are possible to respect.[8] As a direct result of this, transposition and application records have improved in some areas. Attention to technical accuracy and clarity of objectives may, one would think, explain implementation patterns.

Alternatively, international events that either strengthen or weaken the legitimacy and control of the EU could also explain implementation. In 1989 and early 1990, for instance, revolutions in Eastern Europe and German unification pushed Chancellor Helmut Kohl and President François Mitterrand to strengthen the Community's legitimacy and control over national affairs. Part of their strategy included broadening the power and activities of bureaucrats in Brussels. In the 1970s, on the other hand, member states turned protectionist in response to two oil-shocks and inflation and ignored several initiatives and proposals of the EC.

It is impossible to argue against the idea that the 'quality' of a directive, as the first explanation claims, has an inordinate influence over how well it is implemented: the parsimonious and technically sensible nature of some EU directives has undoubtedly eased their implementation. Yet this explanation may actually explain differences in the implementation of different directives. It can say little about *variation* in the implementation of any given directive, however precise or imprecise, across nations. Variation is, however, the dependent variable of this volume.[9] The second explanation suffers from the same problem. It identifies variables (time and legitimacy) that predict differences in the implementation of different directives. The explanation does not address the issue at the heart of this volume: variation in the implementation of the same directive.

From Hewstone's (1986) work on the attitudes of European nations toward the former European Community, one might conclude that public sentiments towards the European Union would affect implementation enormously.[10] At every decision point the views of the public could enter into legislators' considerations and foreign policy choices (Merritt and Puchala 1968).[11] Nations whose European identity is weak assign low priority to any European issue, including directives. Transposition takes time because national and other international issues have precedence. Directives are adapted to national laws in the greatest possible degree to minimize public friction. During the application phase, information and enforcement procedures are kept to a minimum. Sanctions are nonexistent, and the institution building required for application is avoided (Schmitter 1992, 8).

Table A.1 in the appendix shows that this explanation is clearly untenable. There exists an inverse relationship between public disapproval rates and number of infringements. Countries with high public disapproval levels show the fewest infringements, while countries with lower disapproval levels show the most infringements. Public sentiments toward the European Union do not, therefore, have an impact on implementation.

Two more serious explanations could challenge the main thrust of this book. First, political actors may feel favorable or antagonistic toward the EU regardless of public opinion, and may be capable of favoring policies in line with their objectives, *independently* of public sentiments or the challenge of the directives to national institutions.[12] Moravcsik (1994, 1991), building on two-level game theory inhabited by rational actors developed by Putnam (1988), Evans et al. (1993), and others, argues precisely this point. He proposes that the EU increases the independence and control of government party leaders and their close state executives by centralizing power into their hands and freeing their yoke (Moravcsik 1994, 2).

The EU grants more power to "chief executives" and the "head of state or government and his or her closest ministerial supporters" than to any other domestic actor, such as interest groups and opposing political parties (Moravcsik 1994, 3). Their presence, especially in the European Council, redistributes domestic control over institutions, ideas, and information to their advantage. Through them, chief executives control how issues reach parliaments, prepare the nation for their adoption, and mobilize key interest groups and bureaucrats.[13]

Alternatively, one could emphasizes that states are the *means* through which directives are implemented. States are more or less *capable* of transposing and applying directives. Great Britain, for example, with a more efficient and organized legislature and civil service than Italy, is more capable of transposing and applying directives than Italy. Capacity depends on the internal organization of the legislative and executive branches and the efficiency with which they operate (Siedentopf and Hauschild 1988, 2).[14] It is interesting to note that aggregate data lend this explanation apparent validity. Table A.1 in the appendix shows that countries known to have efficient state machineries have the fewest infringements, while countries afflicted by inefficient states have the most infringements (with the notable exception of France).

This book has not addressed these explanations directly. Implicitly, however, it has. Its emphasis on the reactions of political leaders and legislators to the impact that the directives would have had on domestic institutions is intended to 'strip' these actors of the autonomy that the first of these two explanations grants them. With the exception, perhaps, of Mitterand in France and his revitalization of socialist tendencies, in all other cases the behavior of key individuals was *contextualized* as a reaction to the directive's potential threats to domestic institutions and therefore *explained*.

Aggregate data apparently support the second of these explanations. Indeed, countries with generally efficient states have the fewest infringements. The last six chapters, however, consistently pointed out how state efficiency was *not* the determining factor for implementation. Italy's inefficient government was rather quick at adopting the EPD; Great Britain's efficient government was rather slow at adopting the same directive. Spain's inefficient government was fast at adopting the SSD directive. France's fairly efficient government behaved in diametrically opposite fashion when it came to transposition and application of the EPD. Factors other than state efficiency drove implementation, and it is very likely that the apparent relationship between state efficiency and implementation records revealed by table A.1 in the appendix is in fact fortuitous.

Future research should, nevertheless, question the main propositions of this volume. It could concentrate on four types of issues. These issues are intrinsic to the design of this volume. They are: the premise of the explanation (directives target two specific kinds of institutions), the explanation itself (the relationship between a directive and targeted institutions determines implementation), the case studies (two directives in three countries respectively), and the level of analysis (focused on domestic variables).

Several directives do not target national institutions. An overview of such directives and their importance for the EU may challenge the basic premise of the institutional model put forth in this volume. Statistically, for instance, it could be demonstrated that the majority of important directives targets non-institutional variables.

Future research should consider formulating new explanations. The institutional explanation is built on the assumption that directives carry visions that affect interest groups and national policy legacy primarily. Perhaps different premises could lead to more powerful explanatory models. What a directive targets might be in the first instance irrelevant in a number of cases.

Clearly, future research should focus on different directives. There are virtues and faults with trading a high number of case studies for in-depth understanding of events. This volume has only considered two directives and the implementation of each in three countries. In the face of hundreds of directives and over ten countries in the Union for many years, additional evidence should be gathered to test the worth of the findings. One of the major policy areas of the European Union is the economy. This volume has considered directives in the social and environmental spheres. Perhaps rational choice models could successfully explain the implementation of economic directives.

Secondly, although carefully selected against selection bias, France, Great Britain, Spain, and Italy could nevertheless support the institutional model unusually well. Studying the implementation of the same, or different, directives in different countries may weaken or strengthen the institutional argument.

Most importantly, processes outside domestic confines may be, after all, the true determinants of variation in implementation across countries. This volume has examined in detail what happens to a directive once it is enacted and is introduced into member states. Taking the content of a directive as given, the volume has not questioned for instance the impact of the policy-making processes in Brussels on implementation. Assuming for a moment that representatives of one country enjoy special influence over the drafting of a directive, it should be easy to construct an argument about the relationship between influence in Brussels and domestic implementation. Those countries whose interest group representatives, legislators, and administrators (if present) enjoy special influence in Brussels are able to shape the content of the directive in a direction that is least harmful (or, even more likely, beneficial) to domestic institutions. In so doing, they force representatives of countries that would face difficulties in implementation to adopt challenging directives. Their countries would implement the directive well, while other countries would stumble in the process. Arguments such as this are interesting but often problematic as well.

This argument in particular has two major flaws. First, it is not at all an explanation of what drives variation in implementation: it only states that certain representatives will successfully shape the content of directives so that they will be easy to implement, and leaves the question of what determines the ease of implementation aside. Thus, it potentially complements the main argument of this book.[15] Secondly, representation may not mean actual knowledge of what might be beneficial or harmful. There is evidence to suggest precisely that representatives often misunderstand the requests of domestic constituencies. If their actions are in fact misguided, the theory may not even complement that of this book. Additional research could, anyway, test the premise of this argument and determine whether it can in fact complement the institutional explanation offered here.

With the main argument of this book thus still standing, a final interesting question arises. Why do some states agree to directives in the first place, when they know that it will be very difficult to implement them domestically? I will outline here the beginning of an answer, since the question, although interesting, is irrelevant to the book's core problem. First, representatives of countries sign on to directives, but, as noted in the preceding paragraph, they may not be aware of the difficulty of implementation that will arise in their countries. Secondly, representatives at times agree to challenging directives in exchange for the adoption of more beneficial (and easier to implement) directives. Thirdly, not every representative must agree to the directive for it to become official (given that unanimity is no longer required for the adoption of numerous directives, and a 'qualified majority' suffices);[16] states find themselves with directives that they did not support.

Research in all of these areas should shed more light on the variables that affect the implementation of directives and, ultimately, the unification of Europe. It is nevertheless my hope that, in spite of all possible improvements, this volume has in fact afforded important insight into fundamental aspects of the nature and the processes that shape and limit the creation of the European transnational market and that some of its findings may indeed prove relevant for the study of similar efforts in other parts of the world.

Appendix

Table A.1. Infringements of EU Law and Public Sentiments Toward the EU

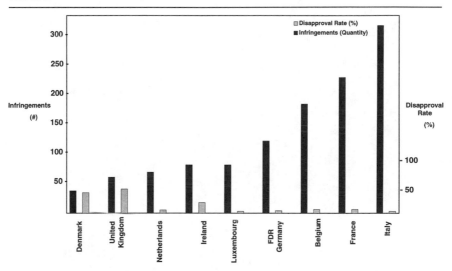

Note: 1. Disapproval Rate based on the question: Do you think that (Respondent's country's) membership in the Common Market is a bad thing? (1973)
2. Infringements are for 1982–1986 and range from Reasoned Opinions to Judgments of the EC Court.
3. The difference in time between infringements and the questionnaire is necessary: infringements reflect violations that occurred typically 1–10 years back.

Source: Data concerning disapproval of the EC from the Commission of the European Communities, *L'Europe vué par les Europeens* (Brussels, 1974).

Notes

Foreword

1. Alan Milward, *The European Rescue of the Nation-State*, Routledge, London, 1992. Cf. Michael Mann, "Has Globalization Ended the Rise and Rise of the Nation-State?" *Review of International Political Economy* 4 (1997), 472–96.

1. Introduction

1. The European Community became the European Union after the Maastricht Treaty of 1993. The treaty laid the groundwork for two additional common policies: security and foreign relations.

2. For a discussion of the poor success rate, see *The Economist*, 22 October 1994, "The European Survey," 16. For a review of the existing literature on EU implementation, see the conclusion below.

3. See Heinz (1991), Keohane and Hoffmann (1990), Leibfried and Pierson (1991), Moravcsik (1991), Nugent (1991), Schmitter (1992), Silvia (1991), Streeck (1993).

4. Jean Monnet, the economic planner of postwar France and Robert Schuman, French foreign minister between 1948 and 1953, were the key players in the design of the Community. Political and economic factors prevented Denmark, Greece, Ireland, Portugal, Spain, and the United Kingdom from joining. The six founding countries "viewed [instead] that the benefits of integration, as opposed to just co-operation, would outweigh what looked to be the major disadvantages—some loss of sovereignty" (Nugent 1991, 26).

5. Three major bodies give birth to the legislation of the European Union. The Commission, composed of representatives from every country serving four-year terms, proposes; the Parliament, composed of representatives from all member states by direct universal suffrage, advises; and the

Council, composed of ministers who alternate according to the nature of the subject being addressed, decides. To these three bodies one must add influential interest groups, international parties, and other official and unofficial factors. The Court, on the other hand, interprets and ensures uniform implementation (Nugent 1991, 54; Ziller 1988, 132).

6. The European Court has repeatedly refused to consider administrative acts as appropriate translations of directives, accepting only laws, for "[administrative acts] by their nature may be altered at the whim of the authorities and lack the appropriate publicity" (Judgements of 25 May 1982, cases 966 and 97/81 (European Court Report (ECR) 1982, 1791, 1819) and of 15 December 1982, case 160/82 (ECR 1982, 463).

7. This stage begins when the executive branch of a government appoints one or more ministries, or creates new ones, for implementation. The choice of ministry varies with the country since the EU makes no recommendations. Ministries in turn select departments, agencies, regions or even cities and request them to draft policies on every aspect of application, including dissemination of information, compliance monitoring, sanctioning, and reporting to EU and national authorities. For a discussion and examples on this, see Siedentopf and Hauschild (1988, 57) and Bennet (1991, 20).

8. Echoing in fact Article 161 (3) of the Treaty on the European Atomic Energy Community and Article 14 (3) of the Treaty on the European Coal and Steel Community.

9. Besides directives, the EU has disposed of regulations and decisions. Regulations take immediate effect after enactment, without need for transposition or administrative mobilization (e.g., the introduction of new speed limits); they are used, however, for less transformative reforms. Decisions are in effect administrative acts rather than legislative acts (e.g., the decision to institute a pilot program) but, like regulations, have limited transformative powers (Nugent 1991, 171; Siedentopf and Ziller 1988).

10. See Addison and Sielbert (1991) on social policy generally; Streek (1993) on the supranational character of social policy; Leibfried and Pierson (1991, 1995) on the pressures and obstacles affecting European social policy; Silvia (1991) and Streeck (1991) on the impact of European policies on labor and trade unions; Hoskyns (1986), Landau (1985), Mazey (1988), and Vogel-Polsky (1985) on the impact of European policies on gender equality.

11. See Ziller (1988, 138) for a brief discussion on the precision of directives. The precision of directives has actually fluctuated throughout time periods depending largely on the extent to which the Community has felt member states should be free and on the legitimacy that the Community has enjoyed: between 1976 and 1979 directives were highly precise, while in the late 1980s an attempt was made "on the initiative of the United Kingdom and

Germany . . . to reduce the density and precision of Community directives" (Siedentopf 1988, 172, 173).

12. See Vogel (1985), Skocpol and Finegold (1983), and Weir (1992) for a general treatment of the endurance of past practices in state structures.

13. The phases of application are: (*a*) selection of ministry, department, or agency responsible for implementation; (*b*) drafting of policies; (*c*) dissemination of information; (*d*) monitoring compliance; (*e*) application of sanctions; and (*f*) filing reports to EU and national authorities.

14. Member states have in most cases authority over the determination of penalties.

Part I: Prologue: The Equal Pay Directive

1. The directive is known as EEC 75/117.

2. At the 1972 Paris summit, Community leaders had in fact endorsed a 'new deal' for European social policy. A specific Social Programme was completed and passed in 1974. It included the Equal Pay Directive (75/117), the Equal Treatment Directive (76/207), and the Social Security Directive (78/7). France had long been concerned that its wage differential, smaller than that of Germany and Great Britain, was hurting its competitiveness internationally. Germany and Great Britain could enjoy lower labor costs because of gender discrimination and women's low wages. See Leibfried and Pierson (1991) and Warner (1984, 142).

3. Even then, women lost a number of cases where Article 119 applied. Consider, for instance, the first of the three famous Defrenne v. Sabena cases (Case 80/70 Defrenne v. Sabena, ECR, 445), where Mrs. Defrenne charged the Belgian national airline, Sabena, of requiring early retirement for women, making them ineligible for social pensions. Although she performed the duties of a steward, her hostess' salary was lower. She lost her case.

4. For an account of earlier attempts by the Commission to develop a more comprehensive law than Article 119, see Warner (1984, 144), Docksey (1987, 7), and Mazey (1988).

2. The French Case: The Importance of the Policy Legacy

1. Comité du Travail Féminin, with representatives from labor, management, and women's organizations (Stetson 1987, 142).

2. The public sector enjoyed already the benefits of equal pay for work of equal value because of the 1946 Civil Service Statute, which rated jobs on a gender-neutral classification system (Lorée 1980, 82-84).

3. See Title IV.

4. The applicative measure for the 1972 act was the decree of 27 March 1973.

5. See Commission of the European Communities (1979, 83–85) for more details on the application procedure.

6. See Lorée (1980, 89) for specific case examples.

7. The minister of social affairs, Joseph Fontanet, admitted during the debates that the fines were too small, but argued against higher fines (Journal Officiel, *Débates Parlementaires*, 21 November 1972, 5562). See Stetson for a discussion of these debates (1987, 143).

8. *Bilan de l'Application de la loi du 22 Décembre 1972*, as referred to in Lorée (1980, 88-90).

9. *Bilan de l'Application de la loi du 22 Décembre 1972*, as referred to in Lorée (1980, 88-90)

10. The Ministère des Droits de La Femme worked for one and a half years on the bill. During that time it consulted numerous times with the Ministries of Labor, Training, Justice, and Treasury, and with representatives from labor, management, and feminist associations (Stetson 1987, 146).

11. The National Assembly, with a Socialist majority, passed the law easily. The Senate proved somewhat more difficult turf, and some "crippling amendments" were introduced (Stetson 1987, 146).

12. Firms were also promised financial assistance if they undertook temporary positive discriminatory policies for women, especially in the areas of training and recruitment. The plan was known as "Vocational Equality Plan."

13. The Roudy Law also banned certain discriminatory practices in hiring and dismissal on the grounds of sex; this however was helpful in implementing the Equal Opportunity Directive of 1976, and not the Equal Pay Directive of 1975.

14. This was in part established by the Auroux Law of 28 October 1982.

15. The Conseil Superieur de l'Égalité.

16. In a poll, taken in December 1983 by the MDF (Citoyennes à Part Entièrie), 93% of "those who had heard of the policy for equality found it important for improving women's equality" . . . "feminist groups too are pleased with the equal opportunity legislation and with the efforts of the [MDF]" (Stetson 1987, 156).

17. See table 10 and tables 1 and 2 in Commission of the European Communities (1979, 98).

18. See Stetson (1987, 146–48), Jenson (1988, 162), Bouillaguet-Bernard and Guavin (1988, 183–84), and Sabourin (1984). This was a period of continuous consultations between government and various social actors concerning the advancement of women. Many well designed initiatives took place, as different branches of government, unions and even management agreed to improve the status of women. See Jenson (1988) and Stetson (1987)

for an account of numerous initiatives, such as educational scholarship programs for girls entering scientific and technical schools, funding for training programs, publicity campaigns, and reforms eliminating sexist images from educational textbooks.

19. The census for every year between 1945 and 1962 shows that the percentage of women in the labor force "was slightly lower than before" (Jenson 1988, 157).

20. See Jenson (1988) for a discussion of trends driving the feminization of the workforce.

21. Married women entering the labor force accounted for much (80%, according to some) of the increase between 1968 and 1975 (Stetson 1987, 129).

22. Chapters 3 and 4 examine the British and Italian turn of events in detail.

23. As suggested by Jenson (1984, 161) and Maruani (1984, 123–24).

24. See Maruani (1984, 123). Women entered primarily the electronic, agribusiness, food, metal-working, chemical, and pharmaceutical industries.

25. They did not occupy the better-paid sectors, such as the chemical, printing and publishing, electrical appliances, and precision equipment sectors. See Boutellier et al. (1975, tables 9 and 10) for a statistical assessment of the destination of women by sector for manual and nonmanual jobs.

26. See Lane (1993, 277–79) for a comparative discussion of the 'gender-regimes' of Germany, France, and Great Britain. She writes: "In France, the gender regime at the level of the state is constituted in a very different manner. The French state has been fairly consistently interventionist, both in its employment and its family policy" (1993, 278). The following pages do not mention that France ratified the Equal Renumeration Convention 100 of the ILO on 10 December 1952. On 25 March 1957, it ratified Article 119 of the Treaty of Rome; indeed, because of France, that article was present in the first place; see Warner (1984, 142–43), Leibfried and Pierson (1991), and Maruani (1984, 134).

27. Numerous explanations exist to account for the interventionist approach of the French state. The weakness of French trade unions in general, and their ambivalence toward women are certainly powerful explanatory variables. Trade unions are discussed in a separate section in this chapter.

28. Secretaire d'État à la Condition Féminine.

29. See Stetson's account (1987, 19, 42–43).

30. More on this committee in the next pages.

31. These legislations followed a rich tradition, later *repealed* in Italy and the United Kingdom, of protective measures dating to the late nineteenth and early twentieth century (see Stetson 1987, 134–36).

32. Journal Officiel, *Débats Parlementaires*, October 1978, 5760, translated by Stetson (1987, 137).

33. By 1975, the proportion of women entering the civil service was at least 50% (see Lorée 1980 and her reference to the Institute National de la Statistique et des Études Économiques, *1975 Census*).

34. Comité du Travail Féminin.

35. Giroud, upon joining the government in 1974 was perhaps nervous about the CTF and opted to seek cooperation rather than conflict (Stetson 1987, 143-44).

36. There is evidence to suggest that Fontanet himself had been charged with the job of designing the bill because of his ambivalent position concerning equality of pay. See Stetson (1987, 143).

37. Those opposed to the bill knew well that, without amendments, application would be very limited. A report produced by management, for instance, had informed them of how the problem of interpreting 'work of equal value' without specific guidelines would discourage enforcement: the Conseil National du Patronat Français had "pointed out in various reports and in the media the difficulties of applying the broader concept of 'equal pay for work of equal value'" and it suggested at a press conference held on 21 October 1975 on 'Femmes et Entreprise' policy guidelines for alternative ways of integrating women into the workplace (Lorée 1980, 105n11).

38. See, for instance, the objections of Socialist Michel Rocard, who argued that the bill "did not provide any means for detecting discriminatory practices" (Lorée 1980, 85), and who argued that it was necessary to tackle the deeper variables that caused pay inequalities, such as training, education, extension of collective agreements, etc. (Stetson 1987, 143).

39. As would have occurred if employers had to raise women's wages.

40. See *Économie et Statistique 81–82* (1976) as quoted in Lorée (1980, 86).

41. The statement was surprising for a minister of a country "where the state has traditionally fine-tuned society through an intricate web of rules and infractions covering all aspects of life" (Stetson 1987, 143); they were less surprising once the prevalent conservative tone of the period is appreciated.

42. The minimum wage was at the time known as SMIG: Salaire Minimum Interprofessional Garanti. Established, by the government, in 1950 by law, it changed later into SMIC: Salaire Minimum Interprofessional de Croissance.

43. Other measures included the shortening of the work week and the lowering of the retirement age.

44. See table 5.4 in Maruani (1984, 124), where the findings of the Organization for Economic Cooperation and Development (*Women and Employment* [Paris, 1980]) are reported.

45. There is no doubt, from the protests of women, unions, and feminists of the time, that the government knew very well that women

primarily would be affected by this strategy. Women already held almost 100% of all existing part-time jobs (a 1975 population census showed this), and on certain occasions the government explicitly stated that women were not the 'breadwinners' of families and should therefore absorb some of the impact of unemployment.

46. Lorée (1980, 99) mentions an official document by the CTF in which the ministry explicitly encourages these measures.

47. While 39.2% of all women between 30 and 34 were working in 1962, 54.6% were working in 1976. All other age groups showed far more modest increases (Institute National de la Statistique et des Études Économiques, *1975 Census*; reported in Maruani 1984, 123).

48. In 1980, Nicole Pasquier became secretary of state for female labor in the Ministry of Labor and appointed a team chaired by Robert Beaudoin. Based on their new report, a bill was presented and adopted by the Council of Ministers. The presidential elections halted the process.

49. Ministère des Droits de la Femme.

50. Specifically, with the Ministries of Labor, Training, Justice, and Treasury.

51. Conseil Superieur de l'Égalité Professional.

52. Specifically, the Ministries of Training and Labor Employment were involved (Stetson 1987, 148).

53. Although there still existed numerous indications that, under Mitterand, traditional viewpoints of female work still existed. Feminists sharply criticized the passage of part-time legislations in 1981 and 1983.

54. In fact, they were the weakest among the unions of eighteen OECD countries considered by Western.

55. In collective agreements, French unions regularly agreed to minimum wages well below the established rates (Rubery and Fagan 1993, 40).

56. Even though, in the case of women, they had clearly recognized the feminization of the workforce (Jenson 1984, 159).

57. See Jenson (1984) for a discussion on this topic.

58. See Jenson (1984, 163), Maruani (1984), and McBride (1985).

59. Including at least eight women in the Bureau Nationale (Jenson 1984, 174).

60. There was a short-lived cooperation between the CFDT and the CGT in the mid-1970s, during which the latter developed a more egalitarian strand, culminating with joint statements on equality. For the complete text of these statements, see CFDT, *Syndacalisme Hebdo* 1 (26 December 1974).

61. See McBride (1985, 50), and "Accord CFDT-CGT sur les Revendications des Femmes Salariées," in *Syndacalisme Hebdo* 1 (26 December 1974).

62. The Communist Party put forth in Parliament bills in the early 1970s that supported, in the words of a party member, the idea that "the status

of a woman is not in fact inseparable from her function in maternity and reproduction" (quoted by Stetson 1987, 133).

63. In the mid-1970s, when it realigned itself with the Socialist Party and espoused more progressive ideas, and in the 1980s, when it collaborated with the Socialists in government, who invited the Communist Party to join. Then, "the party's omnibus proposal emphasized women's rights to work and moved toward the idea of men and women sharing family responsibilities" (Stetson 1987, 133).

64. Had the 1972 act been an alien element in French legislation, transposing the directive would have posed a more serious challenge. As such, the 1972 act stood firmly as one of the major initiatives taken by the French state to regulate the female labor market.

65. As in Italy and Great Britain, respectively.

3. The Italian Case: Strong Union Representation

1. See Ballestrero (1983, 9), Galoppini (1980, 261), and Beccalli (1985, 447).

2. This is not a completely new measure: on 11 July 1969 the Constitutional Court had already opened this possibility to women (*Foro Italiano*, 1969, 1069n123).

3. The law made some exceptions for certain industries, such as manufacturing, unless waived by collective agreements, and for those jobs which must be carried out at night, such as cleaning and security (Article 5).

4. Thus, in 1991, and independently of any EC legislation or pressure, Italy embarked on a positive action program that culminated with important legislative advances, such as Law 125 of 1991.

5. Articles 2 and 3.

6. As obvious from the debates and subsequent letters, such as the circular by the Ministry of Employment itself, in 1978. See Ministero del Lavoro e della Previdenza Sociale (1978).

7. See also Ministero del Lavoro e della Previdenza Sociale (1987) for a series of experts' analyses of the law's impact.

8. Unions have indeed represented workers in most industrialized countries. Wage increases have therefore resulted primarily from union activities (see Cook 1984, 12; International Labour Conference 1985; Bettio 1988; Gregory and Duncan 1981; Zabalza and Tzannatos 1985; and Meulders et al. 1993). Other independent groups and individuals, lacking legitimacy, methods, and coordination have influenced the workplace occasionally (Meehan 1985, 42).

9. In Italy, the old crafts still exist in 1900 but they are overrun by the much larger unions of heterogeneous, unskilled workers that industrialization

creates over a very brief period of time (Beccalli 1984, 188; 1985, 426). Agriculture, moreover, has no antecedent craft union.

10. Leo XII, in the encyclical *Rerum Novarum* of 1891, gave papal sanction to the Catholic labor movement in the fear that women would be recruited by atheist Communist unions.

11. See *Il Lavoro* of 8 January 1907 and other issues as discussed by La Vigna (1985).

12. Notes La Vigna: "Each side became so engrossed in the squabble over whether women were to be saved for the Heaven of Jesus Christ or introduced to the Heaven of the latter-day messiah Karl Marx that they somewhat neglected such earthly union concerns as salary and working conditions" (1985, 130).

13. Which combined in a single text previous laws, specifically 242 of 19 June 1902 and 416 of 7 July 1907.

14. Passages as quoted in La Vigna (1985, 130–150).

15. This claim is made by La Vigna (1985, 149n20) on the bases of reports in the Socialist newspaper *Le Arte Tessili*, noting some as early as in 1901.

16. With regulations: no. 2480 of 9 December 1926, no. 1533 of 4 September 1924, no. 1680 of 28 September 1934, no. 1840 of 16 October 1934, and others.

17. Law 1347 of 5 July 1934 and Law 653 of 26 April 1934.

18. The very first article gives a particular role to the working class: "Italy is a democratic republic founded on labor. Sovereignty belongs to the people, who exercise it within the forms and limits of the Constitution." Only because of a small difference in votes, that same article could have read: "Italy is a workers' republic." See Sassoon (1986, 196–98) for a treatment of the Communist influence on the phrasing of the Constitution.

19. A total of 21, or 4% of the Assembly, were women, a very high percentage for the time (Galoppini 1980, 149). They included Maria Federici and Maria Guidi Cingolani, who were in charge of "the organization of women during the long silence" of the Fascist period (ibid.).

20. At the time national secretary of the textile union, which represented mostly women workers.

21. The collaboration between CGIL and the Communist Party was particularly tight. The collusion of interests occurred often and explicitly: PCI leaders commented on how the party "fulfilled a great part of the activities that should be realized by mass movements organization," while critics noted how the CGIL had "inculcated" in the workers' minds that their most powerful tool was voting, and not strikes (Centro di Studi Sociali e Sindacali 1984, 16). A debate exists over the extent to which leftist parties actually favored women (Ergas 1982, 449; Beccalli 1994).

22. The unusually high percentage of women in Parliament, especially in the PCI, helped coordination; by 1983, 19.2% of PCI representatives to the Chamber of Deputies were women, which, with some other leftist and center parties, brought to more than twice the percentage of women present in the British House of Commons (Sassoon 1986, 107).

23. In 1950 with Law 394 of 23 May 1951.

24. By 1950, the Christian Democrats defected from the CGIL to form the Confederazione Italiana dei Sindacati del Lavoratori (CISL) and the Socialists defected to form the Unione Italiana del Lavoro (UIL). The CGIL remained, and continues today to be, the major union.

25. Consult the series *Annuario di Statistiche del Lavoro* for membership figures. A 1950 supplement put, for 1950, CGIL membership at 5,000,342, CISL membership at 1,489,682, and UIL membership at 401,527.

26. Most European countries, Britain being an exception, acquired such mechanisms over time (Rubery and Fagan 1993, vi).

27. Automatic extension of equal pay collective agreements was never a goal of British trade unions.

28. By 1976, 90% of all workers were covered by collective agreements.

29. There is no reliable data for wage differentials during this time in Italy. Fifteen percent is an educated estimate derived from available estimates for some major sections of industry in this period, specifically those of the EC Commission Report of 1979.

30. Inequality, it is important to underscore, could not be completely eradicated through such measures: first, on average, women in any given sector would earn less because they were less qualified; second, there was always a gap between agreed wage structures and actual wage structures, due to factors such as individual agreements between employers and employees overrunning the original contract, breaches, etc.

31. For a good summary of the objectives of Italian feminism within and outside unions, see Ergas (1980). For a collection of writings on personal experiences, see Bocchio and Torchi (1979). For a collection of original documents concerning meetings, objectives, and strategies of feminist groups, see Frabotta (1976).

32. Milan, Turin, Genoa, and Rome were key locations for the collectives.

33. The first two had ties to the Communist and Christian Democratic Parties, the latter had no official political affiliation.

34. For a thorough report on the activities, themes, organization, and location of these collectives in trade unions but also outside of them, see Calabrò and Grasso (1985).

35. Estimates of participation to these courses between 1972 and 1982 in Milan alone reach 50,000.

36. See Ministero del Lavoro e della Previdenza Sociale (1978) and Confederazione Generale dell'Industria Italiana (1978).

37. Ministero del Lavoro e della Previdenza Sociale (1978).

38. Confederazione Generale dell'Industria Italiana (1978).

39. See Commission of the European Communities (1979, 33); Italy, Luxembourg, and Denmark were the only three countries in which this occurred. Elsewhere, reports, meetings or questions were placed in national Parliaments.

40. As a way out of its responsibilities, the state delegated to unions, in Articles 2 and 3, in very general terms the responsibility of overseeing the application of Law 903.

41. The British Equal Opportunities Commission, created in 1975 following the implementation of the Equal Treatment Directive, served as an example of possible methods for application.

42. A period described by many as one of overproduction in legislative matters and administrative overload.

43. See Ballestrero (1983). Unions especially could have voiced their concerns through the Communist Party.

4. The British Case: The Weak Organization of Working Women

1. The Confederation of British Industry disagreed with the estimates of cost to industry to implement the act, put at first between 3% and 5% of current production costs. Estimating the cost at 6%, it bargained for some time to implement it (Hollingworth 1970; quoted in Soldon 1978, 180).

2. A ruling in 1974 of the European Court restated this: pay means "all emoluments in cash or kind or payable, on condition they are paid, even indirectly, by the employer, as a result of employment" (Case 80/70, *Defrenne v. Sabena*; see ECR 1974, 445 par. 6).

3. Case 12/81, *Garland v. British Railway Engineering*; see European Court Report (ECR) 1982, 359.

4. The *Daily Telegraph* stated at the time that denying employees a right to job evaluation was a "denial of the existence of the right to equal pay" (7 July 1982; quoted in Warner 1984, 153).

5. ECR 1982, 2601, 2606.

6. See Hansard (HL), 16 December 1983, cols. 393–97.

7. See Section 1(3). Material factors would include skill, effort, and decision-making abilities.

8. See Leonard (1987) for a thorough study on the incompetence of judges. According to the findings of the research, only 2.2% of all cases heard by those panels prior to 1975 dealt with discrimination. Of 492 panel persons participating in the 246 sex-discrimination cases heard between 1980 and

1982, 379 individuals were assigned to only one case each, 49 sat on two cases, and 5 on 3. As to the chairpersons, in a sample of 215 cases during that same period with 116 chairpersons, only 7 had heard more than one case a year. The EOC defined tribunals as "industrially illiterate" (Equal Opportunities Commission 1994, 45). Women made up 7% of chairpersons and were found in 21% of panels. In 23% of equal pay cases there were no women in the panels. Only 20% of all cases heard between 1976 and 1983 ended with a victory (Equal Opportunities Commission 1994).

9. See Equal Opportunities Commission (1994).

10. Provided, as already discussed, that the employer had previously failed to defend his practices in a tribunal hearing.

11. Of the 600 claims made mentioned by Hoskyns, only four had, by 1986, gone through the full procedures, and of these three had won and one lost; one of the victories was later nullified on appeal (Hoskyns 1986, 310).

12. See Rubery and Fagan (1993, 40–42) and Lorwin and Boston (1984, 151).

13. The textile industry unions, which, in 1823, had 23% of workers as males. See Hobsawn (1969, 68) as quoted in Lorwin and Boston (1984, 140).

14. Although evidence on strikes, participation, financial aid, and independent initiatives shows the opposite (Soldon 1985, 15). Explicit arguments were made by union officials against pay.

15. These two seats remained two until 1981.

16. Bergamaschi writes: the resolution was a "noteworthy measure since it was approved in a trade union context that aimed, among other things, at controlling access to specialized jobs and that excluded women from acquiring skills and from entering unions themselves" (1989, 134).

17. See Meehan (1985, 22, 37).

18. The league, founded in 1859, sought to oppose the passing of laws limiting women's rights to compete freely in labor markets and argued for employment irrespective of sex.

19. The league's membership, prior to merging with a general union, reached eighty thousand. It imposed nominal fees and established a strike fund. See Holocombe (1973) for a discussion of women's early organizations.

20. Some unions, of course, opposed outright the takeover of women, such as the Cotton Spinners' Amalgamation (Soldon 1978, 92).

21. The Ministry of Munitions issued an instruction ordering employers to give women equal pay if they were doing 'men's work'. This occurred after unions discovered that wages for unskilled and semiskilled positions were lowered without promises for future upgrading. A closer look at the formulation of the instructions shows, however, that women would actually be paid less anyhow (Soldon 1978, 82). During World War II, the TUC and

other independent unions agreed to sign the Extended Employment of Women Agreement, which guaranteed favorable, although not equal, wages.

22. Between 1931 and 1939 participation rates in national textile unions shrank from 282,000 to 182,000.

23. The large National Federation of Women Workers merged with the National Union of General and Municipal Workers.

24. By a margin of 98 to 50 votes (Soldon 1985, 22).

25. Not all unions held such views, of course. Representatives of the Amalgamated Engineering Union, the Union of Shop Distributive and Allied Workers, the Transport and General Workers Union, and others at the Annual Conference of Representatives of Trade Unions Catering for Women Workers (ACUCWW) severely admonished the TUC for its position.

26. See Report of ACUCWW (1947, 7, 30–32), as referred to by Soldon (1978, 159–61).

27. See Report of ACUCWW (1950, 7), as referred to by Soldon (1978, 161).

28. In 1971, for instance, although 59.4% of the National Union of Public Employees were women, only 1 of the 90 full-time officials was a woman (Lorwin and Boston 1984, 140).

29. Mrs. E. Tebbs, of the Society of Graphical & Allied Trades, stated: "although equal pay is not the order of the day, despite Government election promises, I feel we must carry the fight *even* further and demand not just equal pay for equal work but equal pay for work of equal value" (TUC 1965, 415).

30. See *Sun*, 19 April 1965, and the discussion by Soldon (1978, 176).

31. This was heavily criticized by many TUC female members, including Mrs. C. Page of the Union Shop Distributive & Allied Workers, and Mrs. E. A. Hunt of the Association of Scientific Workers (TUC 1965, 415, 416).

32. The strike ended with the dismissal of five thousand workers, an agreement that women would be paid 92% the rate of men and a program for a phased implementation of equal pay. At the Ford Halewood plant in Liverpool two hundred machinists called a sympathy strike. They received the official backing of the Amalgamated Union of Engineering and Foundry Workers Executive (Soldon 1978, 178).

33. In 1968 she unexpectedly replaced Mr. Gunter as minister of labor and was the architect of the EPA bill.

34. The groups included the Six Point Group, the Status of Women Committee, the Suffrage Fellowship, and the British Federation of Business and Professional Women.

35. There is ample evidence to show that the EPA did not come as a result of pressures from feminist groups. See, for instance, Hoskyns (1986, 304).

36. See Cook, Lorwin, and Daniels (1992, 98), Soldon (1985, 26), and TUC (1973, 2,3).

37. There are historical roots to government opposition. Opposition to equal pay because of costs and inflation occurred, for instance, in 1912, when Duff Cooper, parliamentary secretary of the minister of labor, answered negatively to requests of equal pay for equal work forwarded by the Atkins Committee or when government officials opposed equal pay in 1915, arguing that increases in wages to women were unnecessary and unfair to men, who had to support a family with their wages. See Soldon (1978, 93).

38. This is not to say that wage differentials are caused solely by the failure to apply the principle of equal pay for work of equal value. On the contrary, even if this principle is well applied some wage differential is to be expected, as lower qualifications held by women and other variables push women into positions that are qualified as low independently of gender. The point here is that the failure to apply the EPD and the EPA in Britain has precluded any improvement in those areas where jobs are in fact of equal value and has permitted and reinforced discrimination in pay for those jobs traditionally held by women. This failure, unique in the sample of European countries considered here, explains why in part Britain had the largest wage differential in Europe. See Rubery (1992) and Rubery and Fagan (1993) for a discussion of the variables responsible for wage differentials.

39. See also Rubery (1992, 75) and Davies (1987, 27).

40. *Report of the Royal Commission on Equal Pay.*

41. By 1 June 1966, 51 states had ratified Convention 100.

42. In 1963, the Women's Conference published *The Industrial Charter for Women*, in which it asked for equal pay for equal work (Soldon 1978, 174).

43. The quotation is found in a letter to Anne Goodwin, of the TUC Economic Committee, and reported by Soldon (1978, 161).

44. A woman trade union official reacted to this by stating: "Britain's capitalist employers are definitely not the working girl's best friend, and Ray Gunter has been no Prince Charming either" (TUC 1965, 415).

45. See *Daily Telegraph*, 12 July 1967 as discussed by Soldon (1978, 170–75).

46. See Davies (1987, 28).

47. Standing Committee H (HC), 19 March 1970, col. 264; Standing Committee H (HC), 10 March 1970, cols. 264–66; Hansard (HC) 9 February 1970, col. 920.

48. Mrs. Castle, the Fawcett Society, the National Joint Council for Working Women's Organization, the Status of Women Committee, and other groups. At meetings, they pressured the secretary of employment, Chicester-Clark, to step up application procedures and criticized the EPA in published reports (Meehan 1985, 50).

49. Internationally, British officials worked to nullify the directive. They opposed the directive during its formulation. Government representatives fought against the validity of the directive in the second of the famous Defrenne cases, afraid of the significant economic redistribution that could occur. A critic wrote: "Needless to say, both the British and Irish Governments exercised their rights to present observations in the preliminary hearings" (Warner 1984, 149; see also *Financial Times*, 11 March 1976).

50. By Mr. Andrew Bennett MP (Hansard (HC) 20 July 1983, cols. 481–82), Mr. Barry Jones MP (Hansard (HC), 20 July 1983, cols. 487), Ms. Jo Richardson MP (Hansard (HC), 20 July 1983, col. 492), and Mr. Robert Maclennan MP (Hansard (HC) 20 July 1983, col. 497).

51. Writes Hoskyns: the government responded with "extreme ill-will to [the EC] ruling" as evidenced, in Britain, by the contemptuous presentation of the government's amendments in the House of Commons in July 1983 (Hoskyns 1986, 310).

52. Hansard (HC), 20 July 1983, cols. 492, 494–95, and 496–97.

53. See Hansard (HC), 20 July 1983, cols. 690–500.

54. Hansard (HC), 20 July 1983, col. 498.

55. Hansard (HL), 5 December 1983, cols. 882–930.

56. Hansard (HL), 5 December 1983, cols. 882–86.

57. Hansard (HL), 5 December 1983, cols. 901–2.

58. See Hansard (HL), 16 December 1983, cols. 393–97.

59. See Hansard (HL), 4 February 1984, cols. 725–46.

60. See Hansard (HC), 4 February 1984, cols. 359–75.

61. *Equal Pay . . . Making It Work.*

62. See for example the reply by Mr. John Patten MP, on 2 December 1988, Hansard (HC), col. 414; by Mr. Nicholls MP, on 17 January 1989, Hansard (HC), col. 164; by Mr. Jackson MP, on 11 June 1991, Hansard (HC), col. 788; and by Mr. McLoughlin MP, on 6 July 1991, Hansard (HC), col. 47.

63. It is important to note that these were also the years of governmental attacks on labor legislation in general. In 1988, the government published a Consultation Document on Wages Councils (which had ensured minimum wages for industries) recommending their total abolition. In several passages of the document it was implicit that the government knew how the abolition of the 26 remaining Wage Councils would adversely affect women in particular. In paragraph 17 of the document, the government recognized how "the majority of workers [left] are female." Note also how in 1989 the United Kingdom was the only member to vote against the Community Charter of Fundamental Social Rights of 1989.

64. *Report of the United Kingdom and Northern Ireland to the Committee on the Elimination of Discrimination Against Women*, 1991, 77.

65. In 1974, women made up 50% of union membership (Bergamaschi 1989, 139).

66. To the Select Committee on Expenditures, for instance, an engineering union official openly argued that "he did not approve of women working" (Meehan 1985, 40, 41).

67. Craig, Garnsey, and Rubery (1985, 66n83; quoted in Davies 1987, 46).

Part II: Prologue: The SO$_2$ and Suspended Particulates Directive

1. For the sake of brevity, the directive will be referred to hereafter as either the Air Pollution Directive or the SSD.

2. See Johnson and Corcelle (1989, 1), Handler (1994), and Capria (1992, 15). Article 2 indirectly mentions the environment, stating, for instance, that economic expansion must be "balanced" and "harmonious." Article 37 mentions only that the economy should be controlled when it harms human, animal, or vegetable life.

3. The Single European Act, for instance, introduced twenty new articles on the environment in the Treaty of Rome (Vacca 1992, 31–32).

4. See Nugent (1991, 261). See Johnson and Corcelle (1989) and Vacca (1992) for details on the various programs and a list of measures taken. The former correctly argue that the decisive moment for the creation of an environmental policy came around 1983–84, in response to damage done to forests in Germany and other northern countries (Johnson and Corcelle 1989, 110–11).

5. Johnson and Corcelle (1989) note that international pressures stemming from the signing, by the Community and the Soviet Union, of the Geneva Convention on Long-Range Trans-boundary Air Pollution of 1979 explain in part the decision to adopt the directive.

6. Other directives to control air pollution followed. There have been three types of directives in this arena:

> 1. Air quality standards: sulphur dioxide and suspended particulates, lead (1982) and nitrogen dioxide (1985)
> 2. Product quality standards: sulphur content in gas oils (1975) and lead in petrol (1978)
> 3. Clean cars: the eventual elimination of leaded petrol and a number of measures to limit other pollutants.

5. The Italian Case: An Unfavorable Policy Legacy

1. There are 21 regions in Italy. They enjoy some legislative and much administrative autonomy.

2. See, for instance, directives 80/779, 82/884, 84/360, and 85/203, all on air quality standards (Vacca 1992, 159).

3. Articles 11 and 20 of that law delegated the legal authority to ministries for those directives that, because primarily technical or consistent with existing laws, an overburdened Parliament opts to ignore. In practice, a number of very important directives, in legal terms, have been implemented using this alternative (Ciriolo 1991, 69; Vacca 1992, 156).

4. The European Commission thus generally accepts these types of decrees, recognizing that they are substitutes for legal acts.

5. This choice was made in light of the fact that Law 615 of 1966, discussed later, already empowered regions to develop, to their discretion, pollution monitoring networks (Capria 1992, 144).

6. The decree, as noted, did not include winter-median values, related SO_2 and smoke values, and peak values.

7. Decreto Ministeriale 20 May 1991 (Capria 1992, 144).

8. See Bocala (1988).

9. Most of which predated the directive and followed the earlier, rather unsuccessful, pollution law, discussed in the next section: Law 615 of 1966 (Capria 1992, 145). According to various sources, a number of private or community level (i.e., cities) monitoring stations had been established very recently in response to unbearable levels of pollution (especially SO_2) in inhabited centers (Capria 1992, 146). It was now up to the state to coordinate these resources and design a national policy.

10. There exists a number of sources on this matter. See Cederna (1975) for a thorough treatment.

11. See OECD (1979).

12. Farro identifies three types of these groups, from the center to the left of the political spectrum. All shared a vision of more balanced development, with special attention given to the proper growth of urban centers (Farro 1991, 34–37).

13. They organized, for instance, a photographic exhibit in 1967 on the degradation of artistic sites and the publication, in a major national newspaper, of a long series of articles on the subject. See, for example, *Corriere della Sera*, "Ottanta Proposte per Salvare l'Ambiente," 7 April 1972 (as noted in Reich 1984, 388).

14. The former was the most powerful, and one of the oldest conservation groups in Italy (founded in 1955), the latter was a new organization formed in 1972. See Bettini (1984, 145n7) for a profile of some of the most important organizations.

15. The objectives of Italia Nostra were openly recognized as the "protection of the historic, artistic and natural heritage of the country," as

quoted in Rovelli (1988, 77). See Stringher's speech (Accademia Nazionale dei Lincei 1985, 29–31) for a succinct treatment of how Italia Nostra framed its goals in terms of cultural assets during the 1950s, 1960s, and 1970s. The disinterest in nature had cultural roots dating back to the industrialization of the country at the turn of the century. Italians, unlike other Europeans, were fascinated by modernization and ignored its consequences on the natural environment, while seriously considering its implications for art and culture (consider, for instance, the Futurist movement). Italia Nostra organized, for example, a conference in 1970 on the Christian Democrats' management of the city during the previous twenty-five years (Lodi 1988, 19).

16. See Dente and Lewanski (1983, 112, 121) and Cederna (1975) for discussions on this topic.

17. One organization mostly focused on strict ecological problems, without links to the cultural heritage, was the Italian branch of the World Wildlife Fund (WWF). Only in the early 1980s, did a real public consciousness give life to groups solely dedicated to the environment, such as the Lega Ambiente. See Rovelli (1988, 77–78) and Lodi (1988, 23–25) for a discussion of the 1980s and the strengthening of the environmental movement.

18. Between 1980 and 1990, membership for WWF rose from 30,000 to 274,000 (Davidson 1991, 48); membership for the Lega Ambiente rose, between 1983 and 1990, from 15,000 to 50,000, as mentioned in a brochure of the organization in 1990 (Davidson 1991). See also Farro (1991, 284–85), and especially Diani (1988).

19. Some groups insisted on remaining independent from political parties and attacked the government directly in order to retain ideological integrity and to have full control of how issues reached the government. In so doing, they remained reliant primarily on the direct support of individual citizens, at the time still indifferent to their cause (see Rovelli 1988, 91, 94–95). Lodi (1988, 17) notes that some groups, such as the National League for the Protection of Birds, promised political figures votes if they promoted special legislations. Farro notes that the idea of balanced development proposed by intellectual groups was in fact partly received by the government at the time, in particular the Ministry of Economic Programming and Budgeting (Farro 1991, 35n3).

20. We will consider here the views of the two major Italian parties.

21. From an article in *L'Unità*, the PCI's newspaper (13 October 1990, 9-10), as translated in Davidson (1991, 43). See Davidson (1991, 18–44) for an excellent treatment of the PCI's historical antagonism to ecological social movements prior to the late 1980s.

22. Union response paralleled in intensity that of the DC and PCI. Unions took an interest in pollution, but always in relation to how factory work would affect workers' health and how pollution could be understood in

relation to economic modes of production. See the presentation of a trade union officer at a national conference on "Man and His Environment" held in Genova in 1978 (Zingariello 1980, 237). Union actions were, anyhow, of "sporadic occurrence" (Guttieres and Ruffolo 1982, 32).

23. Starting in 1985 and by 1989, all parties in local, regional, national, and European elections included the protection of the environment as part of their agenda (see Farro 1991, 33).

24. Naturally, a number of other factors prevented the development of a proper legal and administrative system for the protection of the environment. Culturally, as mentioned, Italians had exhibited a disinterest in nature and in the consequences of industrialization since the late 1800s, and a fascination with modernization, as exemplified by the Futurist movement. Given that, since World War II it was the somewhat unique (in Europe) failure of social movements and political parties to respond aggressively to the crisis that explains how, as late as 1980, Italy could still have a very backward management of its environment.

25. See Cederna for a general assessment of Italy versus the rest of Europe (1975, 6–92). See also Magi and Imbergamo (1984, 341).

26. With the exception of the outdated, and rather weak, but still valid Law 1497 of 29 June 1939, protecting only 'natural beauties', i.e., selected parts of the territory. Law 1497 was the only law to mention 'nature', according to an expert in 1975 (Cederna 1975, 5, 10). Article 9 of the Constitution underwent revisions until its original attention to the protection of nature as a national patrimony was lost, emphasis was given to the cultural patrimony, and only the importance of protecting the landscape was mentioned. See Cederna (1975, 7–9, 12) for a detailed criticism of Article 9.

27. Later revised with Law 650 of 24 December 1979. From its enactment, the 1976 version was known as a failure. It offered minimal guidelines, as a result of disagreements between regional representatives, and envisioned a system of self-financing that would have never succeeded. The following amendments handed to regions much more authority and provided for more realistic financing (Reich 1984, 385; Guttieres and Ruffolo 1982, 70).

28. On the objectives of the law, see also Capria (1992, 142).

29. The application of the law had actually to wait until 1972, when the necessary presidential decrees specified implementing regulations for three kinds of emitters (DPR 1391 pf. 1970 [thermal plants], 323 of 1971 [diesel engines] and 322 of 1971 [industrial installations]).

30. See Guttieres and Ruffolo (1982, 42, 44–68) for a precise specification of criteria used and targeted emitters.

31. See precise limits in Guttieres and Ruffolo (1982, 53).

32. Guidelines applied to all thermal installations regardless of location, provided they were reasonably powerful (30,000 Kcal per hour).

33. That is, the selection of categories of polluters that would need to be restrained, and the rationale for that selection.

34. See the comprehensive list of pollution laws (including decrees) for all fields in Guttieres and Ruffolo (1982, xvi–xxii).

35. As a result of DPR 616 of 1977, regions enjoyed wide legislative and administrative powers in matters of the environment.

36. Prior to 1972, Article 117 of the Constitution specified a number of areas, including aspects of the environment (for example, 'purity' of water and air), in which regions were allowed to produce legislative measures and expected to have an administrative apparatus for implementation. DPR 616 of 1977 more explicitly handed to regions the responsibility of overseeing the "protection of the environment from pollution" (Article 80). In both the Constitution and DPR 616, however, the state retained authority over the coordination and direction of regional initiatives (Guttieres and Ruffolo 1982, 4–6; see also Maffei 1984, 96).

37. Such as the improved version of the Water Pollution Law of 1976.

38. The northern and central northern regions have produced some laws, while the southern regions have lagged behind (see Guttieres and Ruffolo 1982, xxiii–xxvii).

39. This use, of course, raised a number of questions, outside and inside the courts, concerning the applicability of the article and the nature of the environment as either a collective or private good. For a discussion of these questions, see Maffei (1984, 101–7).

40. See the following cases in the Court of Cassation: 9 March 1979, No. 1463 and 6 October 1979, No. 5172 (mentioned in Guttieres and Ruffolo 1982, 183) and especially Maffei (1984).

41. Law 1497 of 29 June 1939 on the protection of 'natural beauties' mentioned in the previous section assigned responsibility to the Ministry of Education.

42. The comparison made here is based on percentages of GNP spent on the environment and is therefore not based on absolute numbers. In 1972, Italy spent .56% of its GNP on the protection of the environment, compared to an average of 2% to 3% for other European countries (Cederna 1975, 22).

43. See Accademia Nazionale dei Lincei (1985, 6) for the relevant text.

44. See Cederna (1975, 23–24), where he also criticizes Italian universities for having only four chairs (i.e., tenured positions), no laboratories, and almost no courses in ecology. He notes, as well, that in the country there existed only four geologists employed by the government, very few when compared to most countries, including poorer ones, such as Spain and Ghana.

45. Law 5 of 1975, as cited in Reich (1984, 386).

46. See Accademia Nazionale dei Lincei (1985, 5).

47. In 1983, Bettino Craxi, then prime minister, created an ecology ministry, again without portfolio (Reich 1984, 386). The creation of a real ministry had to wait until 1986.

48. Even considering the 1979 amendments to the Water Pollution Law.

49. See Bettini (1984, 150) and Capria (1992, 145).

50. See Accademia Nazionale dei Lincei (1985, 6).

51. Lombardy, for instance, having articulated in the 1970s a fairly serious response to its air pollution problems; it even pressured the state to support its networks and structures by imposing severe requirements on other regions. At the same time, the state had to confront the fact that several regions in the South had developed almost no administrative apparatus.

52. Including provinces and municipalities.

53. As discussed in the "Time and Extent of Transposition and Application" section of this chapter.

54. In Lombardy, for instance, "the whole department of the Milano Engineering Faculty was hired first as consultant and later as a permanent official" (Dente and Lewanski 1983, 115).

55. Emilia Romagna, with relatively low pollution levels, developed an effective *monitoring* system designed to identify areas in which pollution exceeded limits, and pursued only polluters in the most threatened areas. In Lombardy, afflicted by heavy pollution, an *emission-oriented* strategy was pursued (Dente and Lewanski 1983, 115).

56. It is interesting to note how DPR 203/88, besides trying to transpose the directive correctly, also extended the application of Law 615 to the entire territory and to all producers of emissions, thus showing how the government tried to comply with the requirements by relying on *preexistent* methodologies, rather than undertaking new administrative building (Capria 1992, 142, 149).

57. See the protests in Rome in 1983, with 500,000 participants, but especially in 1986 and later, and the proliferation of new associations struggling for the defense of various aspects of the environment (Farro 1991, 63–123).

58. See Farro (1991, 33–34, 64–65, 176–204) on the birth of the Green lists during the elections of 1985 and later, and on the early meetings in Trento where discussions were held for the creation of a Green Party. See also Biorcio (1988) and Rovelli (1988, 91).

59. See Diani (1988) on the increase of membership in Green organizations.

60. In 1988 the ministry had, for the protection of the environment, 870 billion lire at its disposal, or approximately 450 million dollars (Lewanski 1990).

61. Naturally, one of the first efforts made by the minister of the environment, Mario Pavan, was to gain control of the disparate activities and

administrative capacities currently in the hands of all those ministries and bodies previously active in the environment. See his speech at the Accademia Nazionale dei Lincei in the summer of 1987 (Accademia Nazionale dei Lincei 1990).

62. The Galasso Law of 1985.

63. Although of course the problems of the SSD prompted the state for such a law. See note 40 of chapter 2.

6. The British Case: Attempted Distortion of a Legacy

1. As reported in Haigh (1990a, 190).

2. Without a doubt, officials knew well that existing laws could not faithfully transpose or apply the SSD. The report of the Lords' Scrutiny Committee in 1977 on this matter was openly ambivalent on whether compliance was desirable. That report included an excerpt from the Fifth Report of the Royal Commission of 1976 in which it was stated: "we are also opposed to the imposition of air quality standards" (as reported in Haigh, 1990a, 185).

3. The list also included the Pollution Control and Local Government Order of 1978 for Northern Ireland, as reported in Haigh (1990a, 190).

4. As noted and discussed in Haigh (1990a, 190).

5. In 1966, monitoring stations numbered around 1,200. They were coordinated by the Warren Spring Laboratory, under the Department of Trade and Industry and were undergoing a revision at the time the directive was enacted. Circular 11/81 said that the original plan for reduction was being reconsidered. As of April 1988, there were 288 sites, 163 of which were maintained specifically for the EEC (Haigh and Hewett 1991, 85).

6. As a reminder, speed of *application* is operationalized into timely or late application and is measured in light of EC deadlines. Extent of *application* refers to the proportion of objectives, listed in the original directive, that is attained in practice. Objectives are the quantities, deadlines, number of parties, regions, or sectors of industry that the directive seeks to affect.

7. Power stations, controlled by the Central Electricity Generating Board, eliminated their emissions of smoke by using electrostatic filters (Rose 1990, 113). Industrial sites accepted the Clean Air Act and thus reduced their emissions of smoke greatly. The Royal Commission on Environmental Pollution found that industries produced less than 5% of the smoke in the atmosphere.

8. See the values listed by Hughes (1992, 336). See also Rose (1990, 120).

9. In its 1983 letter to the Commission, the Department of Environment listed "twenty-one districts in England, four in Scotland and three in

Northern Ireland which, on the basis of monitoring results over the years, contained areas which would not comply with the limit values by April 1983" (Haigh 1990a, 191). The department noted that action would be undertaken to ensure full compliance by 1993.

10. Standstill principles were not applied because the government deemed them inapplicable to the British case.

11. SI 1989, No. 317, passed under the European Communities Act of 1972.

12. See Hughes (1992, 321–30) for a detailed outline of the act.

13. See Clapp (1994, 43–48) for a brief historical overview.

14. See Clapp (1994, 44–45) for specific concentrations.

15. Some have speculated that, had the fog settled in a different city than London, the reaction would have been much slower and the Clean Air Act may have not been conceived or approved.

16. The second part of this section analyzes the historical context of the Clean Air Act, which in turn will enable us to understand the British resistance to the foreign principles of the directive.

17. The costs of smoke for the population were estimated to be between £150 and 250 million, and took into consideration damages to stone, furniture, clothing, delays to traffic, unburnt fuel, but not health damages. As such, the damages amounted to 1% or 1.5% of the national income (see Clapp 1994, 49).

18. Section 1 of the Clean Air Act of 1956. Dark smoke was defined as smoke "dark or darker than shade 2" on the Ringelmann Chart, and included most industrial and domestic (if involving coal) emissions. See Ashby and Anderson (1977b, figure 7) for an illustration of the Ringelmann Chart.

19. The person responsible for the part of the building where the relevant fireplace is found.

20. For a succinct summary of these exceptions, see Hughes (1992, 331) and Clapp (1994, 50–51).

21. Section 11 of the Clean Air Act of 1956.

22. The Beaver Committee recommended that "the alkali inspectors should assume responsibility for industrial smoke in certain industries such as iron and steel and the potteries" (Clapp 1994, 50).

23. A similiar point will be made for SO_2 and the historical focus on emissions there.

24. Which went beyond black, to include shades of gray according to an objective scale, as already discussed in note 18 of this chapter.

25. That act replaced the ineffective Smoke Nuisance Abatement Act of 1853 and the Sanitary Acts of 1866, which followed from the heroic efforts of W. A. Mackinnon (and his six bills) and John Simon, but could not control increasing pollution. The act did not apply to London until 1891, when the

Public Health (London) Act of 1991 applied to the metropolis the same principles (see Ashby and Anderson 1977a, 10).

26. Interesting excerpts of the law are found in Ashby and Anderson (1977b).

27. It was created under the auspices of the National Health and Kyrle Societies (Ashby and Anderson 1977a, 11–12).

28. The South Kensington Exhibition of 1881–1882.

29. *The Times*, 5, 12, and 26 November 1880 (as quoted, here and hereafter, by Ashby and Anderson 1977a, 13–20).

30. The meeting took place in a prestigious setting and was attended by dukes, professional men, and experts.

31. With the Public Health Act of 1891.

32. Hansard (HC), 30 April 1913, col. 1214. The bill did exempt domestic fires.

33. Hansard (HL), 24 March 1914, col. 671.

34. Before issuing the final report, the committee produced an interim report in which it urged the Central Housing Authority, responsible for approving the construction of a large number of new homes being built for soldiers returning from war, to oppose projects that did not provide for heating systems that used smokeless fuels. The response was minimal. Only after pressures from Lord Newton, the architect of the report, was a copy of the interim report sent to local authorities (Ashby and Anderson 1977b, 197).

35. See *The Times*, 26 March 1923.

36. See Ashby and Anderson (1977b, 199).

37. *The Times*, 11 August 1923.

38. See Hughes (1992, 320–21)

39. Costs of reducing pollution fell upon the polluters themselves, varying naturally from factory to factory.

40. See Vogel (1986, 80–81) for a discussion on this.

41. Quoted and discussed in Sandbach (1982, 74).

42. Quoted in McLaughlin and Foster (1982). As late as the 1980s, the chief inspector was still defending the practice (Haigh 1990a, 175).

43. Quoted in O'Riordan (1979).

44. Industrial production increased between the late 1950s and early 1970s by 60%, while emissions decreased by 64% (Hughes 1992, 321).

45. See Rose (1990, 123).

46. Haigh (1990b) expressed with relief during a conference, following some victories over the CEGB: "One of the great advantages [of electricity privatization] . . . is that the CEGB will no longer be making . . . British air pollution" (as quoted in Rose 1990, 118).

47. In so doing, it received the support of the CBI. Rose (1990) describes in great detail the relationship between the CEGB and the CBI, offering evidence to show that the CEGB worked to convince industrialists of exceptional costs for research and development and alternative methodologies.

48. With the exception of one more attempt at Bankside, later aborted. See the 112th (and last) Report on Alkali etc. Works of 1975, p. 13 (Parliamentary Papers, 1975) as discussed in Clapp (1994). See also Rose (1990, 123).

49. See Rose (1990, 123–31) on FGD and the use of it by Germany, Japan, and the United States. The committee also recognized that SO_2 pollution had for long been in the hands of alkali inspectorates (as discussed in the next pages).

50. The strategy had the perverse effect of causing the transportation of SO_2 directly into Scandinavian countries, of causing great international uproar against Great Britain, and of earning the country the appellative 'the dirty man of Europe'. See Rose (1990, chapter 4) for a thorough treatment of transnational pollution by the United Kingdom.

51. See the section in this chapter on "Time and Extent of Transposition and Application."

52. For some industries, such as iron and steel, the alkali inspectors were charged with controls. By and large, however, local authorities were charged with controlling smoke; by the 1950s, most smoke, after all, was emitted by domestic fires anyway.

53. The Newton's report, for instance, recommended that enforcement be transferred from some 1800 small authorities to 140 more centralized authorities and that the minister of health take the leadership (Ashby and Anderson 1977b, 198).

54. In light of the flexibility needed and historical practices.

55. See Vogel (1986, 81–82) for a discussion of these.

56. Vogel (1986, 79). The 'means' available for curbing pollution were, on the other had, fully described manuals, known as *The Notes on Best Practicable Means*, that the Inspectorate published for every industry (See Vogel 1986, 79–80).

57. See Hughes (1992, 321) for an example of cases.

58. See Vogel (1986, 87–90) for an excellent treatment of the British prosecution model and transgressors, characterized "by the reluctance of the former to prosecute the latter." Prosecution was understood throughout the period as a sign of failure on the part of the inspectorate and company representatives to arrive at a compromise accommodating to both.

59. Quoted in Vogel (1986, 84).

60. See the 1973 Report of the Inspectorates, as quoted in Vogel (1986, 84).

61. Quoted in Rhodes (1981, 149).

62. There is no doubt that the CEGB's strategy was not a solution to the larger problem of SO_2 pollution: besides transnational pollution in Scandinavia, emissions at higher altitudes caused acid rain within the United Kingdom.

63. See Haigh (1990a, 192) and Rose (1990, 119).

64. Royal Commission on Environmental Pollution (1976), as reported in Haigh (1990a, 185).

65. As quoted in Vogel (1986, 77).

66. That is, the government used evidence of clean air.

67. See Haigh (1990a, 190) for a discussion on this.

68. The CEGB's emission policies did not ensure good air at high altitudes. The directive asked, however, for standards at ground level.

69. See Rose (1990, 124–26) for a discussion on this.

70. The CEGB had probably provided exaggerated estimates of costs to the CBI on several occasions (Rose, 1990).

7. The Spanish Case: A Favorable, but Statist, Policy Legacy

1. For the text of the law, consult Ministerio de Justicia (1985, *Disposicion 19376*, pp. 4100–4102). For a reproduction of the text see also Ministerio de Industria y Energía (1990, 117). For a reproduction of the law with references to subsequent legal or tribunal events related to specific articles, see Sánchez Gascón (1994).

2. Local, regional, and national authorities were all entitled to declare zones polluted (Article 6, Section 1). Single individuals could also alert authorities at any level to polluted areas (Article 6).

3. La Comisión Interministerial del Medio Ambiente (CIMA), which was created in 1972 but abolished in 1987.

4. Which higher authorities the plan must reach depended upon who was responsible for declaring the zone polluted. If the government was responsible, then plans had to approved by the Council of Ministers.

5. Royal Decree 833/75, examined later in this chapter.

6. Red Nacional de Vigilancia y Previsión de la Contaminación Atmosferica.

7. Incapable of providing daily measurements, twenty-four hours a day, throughout the year, as needed to apply the directive.

8. The process of installing more automatic stations was taking an unexpectedly long time because the Ministry of Health, in charge of this task, intended "to place the stations in the hands of the autonomous communities which lack the funding needed for modernising and maintaining the network" (Alvarez 1991, 62).

9. Punishments for non-compliance included fines and temporary shutdowns of factories (with special provisions that would guarantee wages to employees) (Ruiz 1994, 145).

10. See Mateo (1992, 333) for details on this case.

11. Up to 1992, ten areas had been in breach of the directive: Gran Bilbao, Madrid , Cartagena, Punta del Sebo, Badalona, Avilés, San Adrián de Besós, Langreo, Cassa de la Selva, and Montcada i Reixach. This information is found in Mateo (1992, 326).

12. The 1972 act had been preceded by the Regulation on Irrigating, Unhealthy, Harmful and Dangerous Activities (approved by decree, 30 November 1961), whose Article 13 required that any activity considered unhealthy due to its pollution of the atmosphere should be controlled, and a national program (the National School for Health) explicitly designed to measure and control air quality be started in specific cities. In 1963 and 1965 Madrid saw plans and actual stations built for SO_2 and smoke. In 1968 Bilbao and Barcelona also built some stations. Regions and town governments were, in addition, able to legislate on environmental matters before and after the 1972 act.

13. For the text of the law, consult Ministerio de Justicia (1972, *Disposición 1885*, pp. 2807–11). For a reproduction of the text see also Ministerio de Industria y Energía (1990, 3–15).

14. Taken from the text of the law, as published by the Ministerio de Industria y Energía (1990, 6).

15. The First Disposition stated: "within one year at the most, the government . . . will dictate the rules for the realization and execution of the present law."

16. Each measure tackled specific sources of pollution. The act and other laws categorized sources of pollution, and it is in accordance with this categorization that the following measures are organized. All these measures were listed in Section 2 of Article 6.

17. Including, besides electric power plants, power plants for industries.

18. See Presidencia del Gobierno (1977, 202).

19. For the text of the law, consult Ministerio de Justicia (1975: *Disposición 8450*, pp. 1059–86). For a reproduction of the text, see also Ministerio de Industria y Energía (1990).

20. Section 2 of Article 29 lists these clean fuels: electric energy, natural gas, oil gases, manufactured gases, and all gases with a low index of SO_2.

21. The government report in 1977 lists these two pollutants as the most important for the network (Presidencia del Gobierno 1977, 781).

22. See Presidencia del Gobierno (1977, 770–90) for an organized and detailed description of the national network.

23. Manual stations could be put to work by less specialized personnel.

24. As of 1977, only the following cities had or were expected to have in a short period of time automatic stations: Madrid, Asturia (Avilés), Barcelona, Bilbao, Coruña (Puentes de García Rodriguez), Huelva, Murcia (Caragena), Sevilla, Tarragona, Valencia, and Valladolid (Presidencia del Gobierno 1977, 786).

25. See Mateo (1992, 321–22) and Presidencia del Gobierno (1977, 785).

26. Most of the readings reported in 1977 reflect monthly statements.

27. By 1991, there were only 3 new automatic stations, although over 10 areas had been found polluted past the acceptable limits in 1977: the stations were in Avilés, Langreo, and Huelva. Five additional stations, managed and financed by city governments, were in place in Cartagena, Barcelona, Madrid, Bilbao, and Sevilla (Alvarez 1991, 62).

28. By 1990, according to one report, only 30% to 40% of existing stations were in a "condition to provide viable information" (Alvarez 1991, 62).

29. See Sden-Diez (1979, 355–58).

30. The Ministries of Industry, of Public Works, and of Health were all charged with duties.

31. See Mateo (1992, 333).

32. As reported in Luke (1992, 13). See also Silvers (1991, 289).

33. See also *Financial Times*, 23 March 1990, "Special Report: Spanish Environment."

34. As the 1977 governmental report could lead one to believe (Presidencia del Gobierno 1977, 203).

35. As it happened in Great Britain, where the state repeatedly commissioned such studies to justify its reluctance to initiate new policies.

36. These included a national commission on the environment and a high council for the environment, put forth respectively by the General Law of the Environment and the Project for Basic Criteria for the Protection of the Environment. See Costa Morata (1985) and Aguilar (1993, 236).

37. The participation of Green groups was practically nonexistent, mostly because these groups hardly existed in the 1970s. Only in the late 1980s and 1990s did Spain witness the surge of a Green movement. See Silvers (1991, 314–15) for recent changes and Varillas and da Cruz (1981) for a review of existing groups in the 1970s.

38. Aguilar speaks of Spanish industry as not "enjoy[ing] institutionalized participation in the policy-making process" (1993, 224). See also Aguilar (1994).

39. See Presidencia del Gobierno (1977, 210).

40. Aguilar notes that "the debate on the implementation deficit has generally assumed that designs promoting co-operation between public and private actors have a better record than those of a hierarchical and exclusive nature" (1993, 226).

41. In a number of other instances, Spain, lacking legal equivalents, did not transpose environmental directives. As of 1993, Spain was the most serious offender of EC environmental law. See Aguilar (1993, figure 1).

42. So writes Silvers (1991, 297), as he reports on the content of a letter sent to Maria Luisa Toribio of the Spanish Greenpeace.

43. For this, Silvers relies on the *Financial Times*, 23 March 1990, "Special Report: Spanish Environment."

44. See Luke (1992, 12).

45. It is in light of this reason that a number of treaties were written in Spain in the late 1980s on the topic of environmental law.

46. The scenario where polluters do not at all cooperate and the administration is capable anyway of applying the law is almost impossible to imagine in a democratic society. Cooperation had to exist.

47. See Silvers (1991, 314–15) for a discussion of the first signs of change.

8. Conclusion: The Future of a United Europe

1. That is, both institutions came under pressure to change.

2. This idea is also consistent with the model: the model specifies how directives will be implemented within the time limits set by the EU, but it says little about what takes place after the deadline has passed. Implementation may still occur according to the model, but belatedly.

3. See Nugent (1991), Hurwitz and Lequesne (1991), Mazey (1988), and Scharpf (1988).

4. See Offe and Wiesenthal (1984), Heinz (1991), Moravcsik (1991), Silvia (1991), and Streeck (1993).

5. See Keohane and Hoffmann (1990), Lindberg and Scheingold (1971), Pinder (1968), and Schmitter (1992).

6. See Siedentopf and Ziller (1988) and Bennett (1991) for two classic examples of 'documentation' of implementation rather than explanation.

7. "A shift in decision-making from political leadership towards respective bureaucracies" has been noted by many (Siedentopf and Hauschild 1988, 75; Siedentopf 1988, 171), to the point, one could almost argue, that "the legislative initiative has passed into the hands of the administrations" (Mayinzt 1984, 175).

8. "The administrations of the Member States" have the "role of intermediary between objectives of Community law and the interests and difficulties of those concerned" (Siedentopf 1988, 171, 175).

9. Yet, the hypothesis should not be fully dismissed. If correct, it can not only explain variation between directives' implementation but, by virtue of this power, also necessarily some aspects of a single directive's implementation.

10. Hewstone's approach differed from Preston's (1994), who analyzed political parties discourse on the EC for the United Kingdom. Hewstone may have used a more authentic indicator of national identity. The attitudes of political parties toward the EC cannot serve as an accurate gauge of a nation's identity, and their effect on implementation is, at any rate, at the heart of the next hypothesis.

11. Also Oskamp (1977) and Feld and Wilgden (1976, 148).

12. See Nugent (1991, 376) and Siedentopf (1988, 173).

13. Moravcsik notes how Edward Heath, British prime minister in 1974, wrote that the primary purpose of the European Council was to permit executives to propose policies, compromises, and issue linkages that ministers, bureaucratic factions, or domestic groups might otherwise block (Heath 1993).

14. See French and Bell (1977), Hall (1986), March and Olsen (1984), Putnam (1986), and Skocpol and Finegold (1983). Countries also vary in the institutions they use to implement EC laws, and attention should therefore be paid to the capacity of these institutions as well. Some EU members, like France, have created central legislative committees and procedures for transposition and application; others, like England, rely mainly on existing institutions, especially the cabinet office (Ciriolo 1991; Ciavarini 1985).

15. Hence, it is a theory that does not conflict with the main explanation of this book, but, if correct, complements it by noting that representatives in Brussels, conscious of the drivers of implementation, shape the content of directives so that their countries will find it easier to implement them.

16. See Vogel (1995: 36) and Nugent (1991, 118–28): the Single European Act abandoned the unanimity rule in favor of qualified majority: fifty-four of seventy-six votes from at least eight members (there were at the time twelve) would be sufficient for adopting new directives.

References

Accademia Nazionale dei Lincei. 1985. *Giornata dell'Ambiente*. Rome: Accademia Nazionale dei Lincei.

———. 1990. *Una Politica per l'Ambiente in Italia: Prospettive e Realizzazioni*. Rome: Accademia Nazionale dei Lincei.

Addison, J. T., and W. S. Siebert. 1991. "The Social Character of the European Community: Evolution and Controversies." *Industrial and Labor Relations Review* 44 (4): 597–625.

Aguilar, Susana. 1993. "Corporatist and Statist Designs in Environmental Policy: The Contrasting Roles of Germany and Spain in the European Community Scenario." *Environmental Politics* 2 (2): 223–47.

———. 1994. "Convergence in Environmental Policy? The Resilience of National Institutional Designs in Spain and Germany." *Journal of Public Policy* 14 (1): 39–56.

Alvarez, Cristina. 1991. "Sulphur Dioxide and Suspended Particulates: Spain." In Bennett, ed., pp. 60–65.

Amendola, Gianfranco, and Claudio Botre. 1978. *Italia Inquinata*. Rome: Editori Riuniti.

Anales de Moral Social Y Economica. 1979. *Ecologia Y Medio Ambiente*. Madrid: Centro de Estudios Sociales del Valle de los Caidos.

Ashby, E., and Mary Anderson. 1977a. "Studies in the Politics of Environmental Protection: The Historical Roots of the British Clean Air Act, 1956: II. The Appeal to Public Opinion over Domestic Smoke, 1880–1982." *Interdisciplinary Science Reviews* 2 (1): 9–26.

———. 1977b. "Studies in the Politics of Environmental Protection: The Historical Roots of the British Clean Air Act, 1956: III. The Reopening of Public Opinion, 1892-1952." *Interdisciplinary Science Reviews* 2 (3): 191–207.

Ballestrero, Maria Vittoria. 1979. *Dalla Tutela alla Parità: La Legislazione Italiana sul Lavoro delle Donne*. Bologna: Il Mulino.

———. 1983. "Legge di Parità e Disciminazione del Lavoro Femminile." In Ballestrero et al., eds., pp. 9–47.

Ballestrero, Maria Vittoria, Renato Livraghi, Luigi Frey, and Gian Claudio Mariani, eds. 1983. *Lavoro Femminile, Formazione e Parità Uomo-Donna*. Milan: Franco Angeli Editore.

Battiston, Lea, and Gianna Gilardi. 1992. *La Parità tra Consenso e Conflitto: Il Lavoro delle Donne dalla Tutela alle Pari Opportunità, alle Azioni Positive*. Rome: Ediesse.

Beccalli, Bianca. 1984. "Italy." In Cook et al., eds., pp. 184–214.

———. 1985. "Le Politiche del Lavoro Femminile in Italia: Donne, Sindacati e Stato tra il 1974 e il 1984." *Stato e Mercato* 15 (December): 423–59.

———. 1994. "The Modern Women's Movement in Italy." *New Left Review* 204 (March and April): 86–110.

Benedizione, Nicola. 1988. "Rappresentanza di Interessi, Rappresentanza Politica e Tutela dell'Ambiente." In Dente, ed., pp. 70–100.

Bennett, Graham, ed. 1991. *Air Pollution Control in the European Community: Implementation of the EC Directives in the Twelve Member States*. London: Graham and Trotman.

Bergamaschi, Myriam. 1989. "Sindacato, Parità Salariale e Pari Opportunità in Gran Bretagna." In Frey et al., eds., pp. 133–68.

Berger, Suzanne, ed. 1981. *Organizing Interests in Western Europe*. Cambridge: Cambridge University Press.

Bettini, Romano. 1984. "Problemi e Prospettive di Partecipazione Amministrativa nell'Introduzione della 'Via' in Italia." In Greco, ed., pp. 142–59.

Bettio, Francesca. 1988. *The Sexual Division of Wage Labour: The Italian Case*. Oxford: Oxford University Press.

Biorcio, Roberto. 1988. "Ecologia Politica e Liste Verdi." In Biorcio and Lodi, eds., pp. 113–45.

Biorcio, Roberto, and Giovanni Lodi, eds. 1988. *La Sfida Verde: il Movimento Ecologista in Italia*. Padua: Liviana Editrice.

Bocala, W. 1988. *L'Inquinamento Atmosferico in Italia*. Rome: Ente per le Nuove Tecnologie, l'Energia e l'Ambiente (ENEA).

Bocchio, Flora, and Antonia Torchi. 1979. *L'Acqua in Bocca*. Milan: La Salamandra.

Bonifazi, Alberto, and Gianni Salvarani. 1976. *Dalla Parte dei Lavoratori*, vol. 3. Milan: Franco Angeli.

Borgogelli, Franca. 1987. *Il Lavoro Femminile tra Legge e Contrattazione*. Milan: Franco Angeli.

Bouillaguet-Bernard, Patricia, and Annie Gauvin. 1988. "Women, the State and the Family in France: Contradictions of State Policy for Women's Employment." In Rubery, ed., pp. 162–90.

Bouteiller, Jacques, Barrie O. Pettman, and Roger A. Beattie. 1975. *Male and Female Wage Differentials in France: Theory and Measurement.* Patrington (Hull), U.K.: International Institute of Social Economics.

Brewster, Chris, and Paul Teague. 1989. *European Social Policy: Its Impact on the UK.* London: Institute of Personnel Management.

Calabrò, Anna, and Laura Grasso, eds. 1985. *Dal Movimento Femminista al Movimento Diffuso.* Milan: Franco Angeli.

Capotorti, Francesco. 1988. "Legal Problems of Directives, Regulations and their Implementation." In Siedentopf, Heinrich, and Ziller, eds., pp. 151–68.

Capria, Antonella. 1991. "Sulphur Dioxide and Suspended Particulates: Italy." In Bennett, ed., pp. 69–70.

———. 1992. *Direttive Ambientali CEE e Stato di Attuazione in Italia.* Milan: Giuffré Editore.

Cederna, Antonio. 1975. *La Distruzione della Natura in Italia.* Turin: Einaudi.

Centro di Studi Sociali e Sindacali. 1984. *Sindacalisti in Parlamento,* vol. 3. Rome: Ediesse.

Ciavarini Azzi, Giuseppe, ed. 1985. *The Implementation of Community Law by the Member States.* Maastricht, Netherlands: European Institute of Public Administration.

Ciriolo, Antonio. 1991. *Il Dipartimento per il Coordinamento delle Politiche Comunitarie: L'Applicazione delle Direttive CEE in Italia.* Lecce, Italy: Milella.

Clapp, B. W. 1994. *An Environmental History of Britain since the Industrial Revolution.* New York: Longman.

Commission of the European Communities. 1979. *Report of the Commission to the Council on the Application at 12 February 1978 of the Principle of Equal Pay for Men and Women.* Brussels: Commission of the European Communities.

Confederazione Generale dell'Industria Italiana. 23 January 1978. *Protocollo N. 2590/169930.* Rome: Confederazione Generale dell'Industria Italiana.

Cook, Alice. 1984. "Introduction." In Cook et al., eds., pp. 3–36.

Cook, Alice, Val Lorwin, and Arlene Kaplan Daniels, eds. 1984. *Women and Trade Unions in Eleven Industrialized Countries.* Philadelphia: Temple University Press.

———. 1992. *The Most Difficult Revolution: Women and Trade Unions.* Ithaca, N.Y.: Cornell University Press.

Costa Morata, Pedro. 1985. *Hacia la Destrucción Ecológica de España.* Barcelona: Grijalbo.

Craig, C., E. Garnsey, and J. Rubery. 1985. *Pay Structures and Smaller Firms: Women's Employment in Segmented Labour Markets.* London: Department of Employment.

Davidson, Christopher Dmitri. 1991. "Old and New Politics in Italy: the Communist Party and the Environmentalist Movement." Undergraduate thesis, Harvard University, Cambridge, Mass.

Davies, P. L. 1987. "European Equality Legislation, U.K., Legislative Policies and Industrial Relations." In McCrudden, ed., pp. 23–51.

De Paz, M., P. Maifredi, M. Pilo, and M. T. Torti, eds. 1978. *L'Uomo e il Suo Ambiente: Economia delle Risorse Umane e Naturali (Primo Convegno Nationale, Genova 23–25 Novembre 1978)*. Milan: Franco Angeli.

Dente, Bruno, ed. 1990. *Le Politiche Pubbliche in Italia*. Bologna: Il Mulino.

Dente, Bruno and Rudy Lewanski. 1983. "Implementing Air Pollution Control in Italy: The Importance of the Political and Administrative Structure." In Downing and Hanf, eds., pp. 107–28.

Department of Environment, Central Unit of Environmental Pollution. 1976. *Pollution Control in Britain: How It Works*. London: Her Majesty's Stationery Office.

Diani, Mario. 1988. *Isole nell'Arcipelago*. Bologna: Società Editrice il Mulino.

Dockesy, Christopher. 1987. "The European Community and the Promotion of Equality." In McCrudden, ed., pp. 1–22.

Downing, Paul B., and Kennenth Hanf, eds. 1983. *International Comparisons in Implementing Pollution Laws*. Boston: Kluwer-Nijhoff Publishing.

Equal Opportunities Commission. 1993. "Request by the Equal Opportunities Commission to the European Commission to Clarify Whether British Equal Pay Laws Comply with EC Legislation on Equal Pay." Unpublished report.

———. 1994. "Request to the Commission of the EC by the Equal Opportunities Commission for Great Britain in Relation to the Implementation of the Principle of Equal Pay." Unpublished report.

Ergas, Yasemin. 1980. "Femminismo e Crisi di Sistema: Il Percorso Politico delle Donne Attraverso gli Anni Settanta." *Rassegna Italiana di Sociologia* 4 (October and December): 543-568.

———. 1982. "Allargamento della Cittadinanza e Governo del Conflitto: le Politiche Sociali negli Anni Settanta in Italia." *Stato e Mercato* 6 (December): 429-63.

Eurostat Review. 1980. *1970–1979*. Luxembourg: Statistical Office of the European Communities.

Evans, Peter, Dietrich Rueschemeyer, and Theda Skocpol, eds. 1985. *Bringing the State Back In*. Cambridge: Cambridge University Press.

Evans, Peter B., Harold K. Jacobson, and Robert D. Putnam, eds. 1993. *Double-Edged Diplomacy: International Bargaining and Domestic Politics*. Berkeley: University of California Press.

Farro, Antimo. 1991. *La Lente Verde: Cultura, Politica e Azione Ambientalistiche*. Milan: Franco Angelo.

Feld, Werner J., and John K. Wildgen. 1976. *Domestic Political Realities and European Unification: A Study of Mass Publics and Elites in the European Community Countries.* Boulder, Colo.: Westview Press.

Frabotta, Biancamaria, ed. 1976. *La Politica del Femminismo 1973–76.* Rome: Savelli.

French, Wendell L., and Cecil H. Bell. 1977. *Organisationsentwicklung.* Bern and Stuttgart: Haupt.

Frey, Luigi, Renata Livraghi, Tiziano Treu, Maurella Zerbini, and Myriam Bergamaschi, eds. 1989. *Comparable Worth e Segregazione del Lavoro Femminile.* Milan: Franco Angeli.

Gaeta, Lorenzo, and Iolanda Jannelli. 1993. "Il Comitato Nazionale per la Parità." In Gaeta and Zoppoli, eds., pp. 132–52.

Gaeta, Lorenzo, and Lorenzo Zoppoli, eds. 1993. *Il Diritto Diseguale: La Legge sulle Azioni Positive.* Turin: G. Giappichelli Editore.

Galoppini, Annamaria. 1980. *Il Lungo Viaggio Verso la Parità.* Bologna: Zanichelli.

George, Stephen. 1985. *Politics and Policies in the European Community.* Oxford: Clarendon Press.

Goldstein, Judith. 1986. "The Political Economy of Trade: Institutions of Protection." *American Political Science Review* 80 (1): 161–84.

Greco, Nicola, ed. 1984. *La Valutazione di Impatto Ambientale: Rivoluzione o Complicazione Amministrativa?* Milan: Franco Angeli.

Gregory, R. G., and R. C. Duncan. 1981. "Segmented Labour Market Theories and the Australian Experience of Equal Pay for Women." *Journal of Post-Keynesian Economics* 3 (Spring): 403–28.

Guttieres, Mario, and Ugo Ruffolo, eds. 1982. *The Law and Practice Relating to Pollution Control in Italy.* London: Graham and Trotman.

Haigh, Nigel. 1990a. *EEC Environmental Policy and Britain.* Harlow, U.K.: Longman.

———. 1990b. "The Overall European Scene." Paper presented at the Conference on Desulphurization in Coal Combustion Systems, Institution of Chemical Engineers, University of Sheffield, April 1990.

Haigh, Nigel, and Jonathan Hewett. 1991. "Sulphur Dioxide and Suspended Particulates: the United Kingdom." In Bennett, ed., pp. 80–85.

Hall, Peter A. 1986. *Governing the Economy: The Politics of State Intervention in Britain and France.* New York: Oxford University Press.

Handler, Thomas, ed. 1994. *Regulating the European Environment.* Chichester, U.K.: Chancery Law Publishing.

Heath, Edward. 1993. "At the Heart of Europe." In Price, ed., pp. 205–28.

Heinz, Waltzer, ed. 1991. *Status Passages, Institutions and Gatekeeping.* Weinheim, Germany: Deutscher Studienverlag.

Hepple, Bob. 1987. "The Judicial Process in Claims for Equal Pay and Equal Treatment in the United Kingdom." In McCrudden, ed., pp. 143–28.

Hewstone, Miles. 1986. *Understanding Attitudes to the European Community: A Social Psychological Study in Four Member States*. Cambridge: Cambridge University Press.

Hill, Michael. 1983. "Air Pollution Control in Britain: A Case Study in Central-Local Government Reactions." Unpublished paper.

Hobsawn, E.J. 1969. *Industry and Empire*. London: Penguin Books.

Hollingworth, Claire. 1970. "Equal Pay—and Its Price." *Daily Telegraph*, 2 February.

Holocombe, Lee. 1973. *Victorian Ladies at Work: Middle-Class Working Women in England and Wales, 1850–1914*. Hamden, Conn.: Archon Books.

Hoskyins, Catherine. 1986. "Women, European Law and Transnational Politics." *International Journal of the Sociology of Law* 14: 299–315.

Hughes, David. 1992. *Environmental Law*. London: Butterworths.

Hurwitz, Leon, and Christian Lequesne, eds. 1991. *The State of the European Community: Policies, Institutions & Debates in the Transition Years*. Boulder, Colo.: Lynne Rienner Publishers.

International Labor Conference. 1985. *71st Session*. Geneva, Switzerland: International Labour Office.

Jenson, Jane. 1984. "The 'Problem' of Women." In Kesselman, ed., pp. 159–75.

———. 1988. "The Limits of 'and the' Discourse." In Jenson et al., eds., pp. 155–72.

Jenson, Jane, Elizabeth Hagen, and Reddy Ceallaigh, eds. 1988. *Feminization of the Labor Force: Promises and Paradoxes*. New York: Oxford University Press.

Johnson, Stanley P., and Guy Corcelle. 1989. *The Environmental Policy of the European Communities*. London: Graham and Trotman.

Katzenstein, Peter. 1985. *Small States in World Markets: Industrial Policy in Europe*. Ithaca, N.Y.: Cornell University Press.

Keohane, Robert O., and Stanley Hoffmann. 1990. "European Community Politics and Institutional Change." Working Paper No. 25, Minda de Gunzburg Center for European Studies, Harvard University.

Keohane, Robert O., and Helen V. Milner, eds. 1996. *Internationalisation and Domestic Politics*. Cambridge: Cambridge University Press.

Kesselman, Mark. 1984. "Introduction: The French Workers' Movement at the Crossroads." In Kesselman, ed., pp. 1–14.

———, ed. 1984. *The French Workers' Movement: Economic Crisis and Political Change*. London: Allen & Unwin.

Landau, Eve. 1985. *The Rights of Working Women*. Luxembourg: Commission, European Perspective.

Lane, Christel. 1993. "Gender and the Labour Market in Europe: Britain, Germany and France Compared." *The Sociological Review* 42 (2): 274–301.

LaVigna, Claire. 1985. "Ideology and Trade Unions: A Study of Italian Working Women." In Soldon, ed., pp. 125–52.

Lawrence, Elizabeth. 1994. *Gender and Trade Unions*. London: Taylor and Francis.

Leibfried, Stephan, and Paul Pierson, eds. 1991. "Emergent Supranational Policy: The EC Social Dimension in Comparative Perspective." Draft, Center for European Studies, Harvard University.

———, eds. 1995. *European Social Policy: Between Fragmentation and Integration*. Washington, D.C.: The Brookings Institution.

Leonard, Alice. 1987. *Judging Inequality: The Effectiveness of the Industrial Tribunal System in Sex Discrimination and Equal Pay Cases*. London: Cobden Trust.

Lewanski, Rodolfo. 1990. "La Politica Ambientale." In Dente, ed., pp. 281–314.

Lindberg, Leon, and Stuart Scheingold. 1971. *Regional Integration: Theory and Research*. Cambridge, Mass.: Harvard University Press.

Linz, Juan J. 1981. "A Century of Politics and Interests in Spain." In Berger, ed., pp. 356–415.

Lodi, Giovanni. 1988. "L'Azione Ecologica in Italia: dal Protezionismo Storico alle Liste Verdi." In Biorcio and Lodi, eds., pp. 17–26.

Lorée, Marguerite. 1980. "Equal Pay and Equal Opportunity Law in France." In Steinberg, ed., pp. 79–107.

Lorwin, Val, and Judith Boston. 1984. "Great Britain." In Cook et al., eds., pp. 140–61.

Luke, Anthony. 1992. "Spain: Too Poor to Be Green." *New Scientist* 135 (25 July): 12–13.

Maffei, Corrado. 1984. "Valutazione d'Impatto Ambientale e Documentazione Giuridica per la Protezione degli Interessi Diffusi: Tradizione ed Innovazione." In Greco, ed., pp. 64–140.

Magi, Franco, and Giovanni Imbergamo. 1984. "Un'Esperienza di Impatto Ambientale negli Anni '70: Tentativo di Confronto con la Direttiva CEE in Fase di Elaborazione." In Greco, ed., pp. 327–46.

Mateo, Ramón Martín. 1992. *Tratado de Derecho Ambiental*, vol. 2. Madrid: Editorial Trivium.

March, James, and Johan Olsen. 1984. "The New Institutionalism: Organizational Factors in Political Life." *American Political Science Review* 78 (3): 734–49.

Maruani, Margaret. 1984. "France." In Cook et al., eds., pp. 120–39.

Mayintz, Renate. 1984. "German Federal Bureaucrats: A Functional Elite between Politics and Administration." In Suleiman, ed., pp. 174–205.

Mazey, Sonia. 1988. "European Community Action on Behalf of Women: The Limits of Legislation." *Journal of Common Market Studies* 27 (1): 64–84.

McBride, Theresa. 1985. "French Women and Trade Unionism: The First Hundred Years." In Soldon, ed., pp. 35–50.

McCrudden, Christopher, ed. 1987. *Women, Employment and European Equality Law*. London: Eclipse Publications.

McLaughlin, J., and M. J. Forest. 1982. *The Law and Practice Relating to Pollution Control in the United Kingdom*. London: Graham and Trotman.

Meehan, Elizabeth, 1985. *Women's Rights at Work: Campaigns and Policy in Britain and the United States*. London: Macmillan.

Mény, Yves, Pierre Muller, and Jean-Louis Quermonne, eds. 1996. *Adjusting to Europe: The Impact of the European Union on National Institutions and Policies*. New York: Routledge.

Merritt, R. L., and D. J. Puchala, eds. 1968. *Western European Perspectives on International Affairs*. New York: Praeger.

Meulders, Daniele, Robert Plasman, and Valerie Vander Stricht. 1993. *The Position of Women in the Labour Market in the European Community*. Aldershot, U.K.: Dartmouth Publishing Company.

Ministerio de Justicia. 1972. *Boletín Official del Estado: Disposiciones Generales: 1972*. Madrid: Ministerio de Justicia.

———. 1975. *Boletín Official del Estado: Disposiciones Generales: 1975*. Madrid: Ministerio de Justicia.

———. 1985. *Boletín Official del Estado: Disposiciones Generales: 1985*. Madrid: Ministerio de Justicia.

Ministerio de Industria y Energía. 1990. *Protección del Ambiente Atmosférico*. Madrid: Centro de Publicaciones.

Ministero del Lavoro e della Previdenza Sociale. 1978. *Circular No. 92/78*. Rome: Ministero del Lavoro e della Previdenza Sociale.

———. 1985. *Relazione sullo Stato di Attuazione della Legge 9 Dicembre 1977, No. 903, Anni 1981 e 1982*. Rome: Ministero del Lavoro e della Previdenza Sociale.

———. 1987. *Donne e Lavoro: Analisi e Proposte*. Rome: Ministero del Lavoro e della Previdenza Sociale.

Moravcsik, Andrew. 1991. "Negotiating the Single European Act: National Interests and Conventional Statecraft in the European Community." *International Organization* 45: 19–56.

———. 1994. "Why the European Community Strengthens the State: Domestic Politics and International Cooperation." Paper presented at the Conference of Europeanists, Chicago, April 1994.

Nugent, Neill. 1991. *The Government and Politics of the European Community*. Durham, N.C.: Duke University Press.

Offe, Claus, ed. 1985. *Disorganized Capitalism*. Cambridge, Mass.: MIT Press.

Offe, Claus, and Helmut Wiesenthal. 1985. "Two Logics of Collective Actions." In Offe, ed., pp. 170–220.

Organisation for Economic Cooperation and Development (OECD). 1979. *The State of the Environment in OECD Member Countries.* Paris: OECD.

O'Riordan, Timothy. 1979. "The Role of Environmental Quality Objectives: The Politics of Pollution Control." In O'Riordan and D'Arge, eds., pp. 39–53.

O'Riordan, Timothy, and Ralph C. D'Arge, eds. 1979. *Progress in Resource Management and Environmental Planning,* vol. 1. New York: Wiley.

Oskamp, Stuart. 1977. *Attitudes and Opinions.* Englewood Cliffs, N.J.: Prentice Hall.

Pierson, Paul. 1996. "The Path to European Integration: A Historical Institutional Analysis." *Comparative Political Studies* 29: 123–63.

Pinder, John. 1968. "Positive Integration and Negative Integration: Some Problems of Economic Union in the EEC." *World Today* 24: 88–110.

Presidencia del Gobierno. 1977. *Medio Ambiente en España: Informe General.* Madrid: Subsecretaria de Planificación.

Preston, Peter. 1994. *Europe, Democracy, and the Dissolution of Britain: An Essay on the Issue of Europe in UK Public Discourse.* Brookfield, Vt.: Dartmouth Publishing Company.

Price, Geraldine, ed. 1993. *A Window on Europe: The Lothian European Lectures 1992.* Edinburgh: Canongate Press.

Putnam, Robert. 1986. *The Comparative Study of Political Elites.* Englewood Cliffs, N.J.: Princeton Hall.

———. 1988. "Diplomacy and Domestic Policy." *International Organization* 42: 427–60.

Reich, Michael R. 1984. "Mobilizing for Environmental Policy in Italy and Japan." *Comparative Politics* 16 (1): 379–402.

Rhodes, Gerald. 1981. *Inspectorates in British Government.* London: Allen & Unwin.

Rose, Chris. 1990. *The Dirty Man of Europe: The Great British Pollution Scandal.* London: Simon & Schuster.

Rovelli, Cesare. 1988. "I Modelli Organizzativi delle Associazioni Ambientaliste." In Biorcio and Lodi, eds., pp. 73–98.

Royal Commission on Environmental Pollution. 1976. *Air Pollution Control: An Integrated Approach (Fifth Report).* London: Her Majesty's Stationery Office.

Rubery, Jill. 1992. *The Economics of Equal Value.* Manchester, U.K.: Equal Opportunities Commission.

Rubery, Jill, and Colette Fagan. 1993. *Wage Determination and Sex Segregation in Employment in the European Community.* Report for the Equal Opportunities Unit, Commission of the European Communities.

Ruiz, Xavier. 1994. "Environmental Regulation in Spain." In Handler, ed., pp. 143–56.

Rubery, Jill, ed. 1988. *Women and Recession*. London: Routledge & Kegan Paul.

Sabourin, Annie. 1984. *Le Travail des Femmes dans la CEE e les Conditions Juridiques*. Paris: Economica.

Sánchez Gascón, Alonso. 1994. *Código de las Leyes del Medio Ambiente: Legislación Estatal y Jurisprudencia*. Pamplona, Spain: Aranzadi.

Sandbach, Francis. 1982. *Principles of Pollution Control*. London: Longman.

Sassoon, Donald. 1986. *Contemporary Italy: Politics, Economy and Society since 1945*. New York: Longman.

Scharpf, Fritz J. 1988. "The Joint-Decision Trap: Lessons from German Federalism and European Integration." *Public Administration* 66 (3): 239–78.

Schmitter, Philippe C. 1992. "Interests, Powers and Functions: Emergent Properties and Unintended Consequences in the European Polity." Paper presented at the Conference on the Future of the Euro-Polity, Center for Advanced Studies in the Behavioral Sciences, Palo Alto, 22–25 May.

Sden-Diez, Juan Ignacio. 1979. "Protección del Medio Ambiente Ante la Ley." In Anales de Moral Social Y Economica, pp. 351–59.

Siedentopf, Heinrich. 1988. "The Implementation of Directives in the Member States." In Siedentopf and Ziller, eds., pp. 169–80.

Siedentopf, Heinrich, and Christoph Hauschild. 1988. "The Implementation of Community Legislation by the Member States: A Comparative Analysis." In Siedentopf and Ziller, eds., pp. 1–87.

Siedentopf, Heinrich, and Jacques Ziller, eds. 1988. *Making European Policies Work: The Implementation of Community Legislation in the Member States*. London: Sage.

Smith, Harold. 1981. "The Problem of Equal Pay for Equal Work in Great Britain during World War II." *Journal of Modern History* 53 (4): 653–72.

Silvers, George Matthew. 1991. "The Natural Environment in Spain: A Study of Environmental History, Legislation and Attitudes." *Tulane Environmental Law Journal* 5 (1): 285–316.

Silvia, Steven J. 1991. "The Social Character of the European Community: A Defeat for European Labor." *Industrial and Labor Relations Review* 44 (4): 626-34.

Skocpol, Theda, and Kennenth Finegold. 1983. "State Capacity and Economic Intervention in the Early New Deal." *Political Science Quarterly* 97: 255–78.

Soldon, Norbert C. 1978. *Women in British Trade Unions: 1874–1976*. Dublin: Gill and Macmillan.

————. 1985. "British Women and Trade Unionism: Opportunities Made and Missed." In Soldon, ed., pp. 11–33.

————. ed. 1985. *The Word of Women's Trade Unionism: Comparative Historical Essays*. Westport, Conn.: Greenwood Press.

Steinberg, Ronnie. 1980. *Equal Employment Policy for Women: Strategies for Implementation in the United States, Canada, and Western Europe*. Philadelphia: Temple University Press.

Stetson, Dorothy McBride. 1987. *Women's Rights in France*. Westport, Conn.: Greenwood Press.

Streeck, Wolfgang. 1991. "More Uncertainties: West German Unions Facing 1992." *Industrial Relations* 30 (3): 317–49.

Streeck, Wolfgang. 1993. *From Market-Making to State-Building? Reflections on the Political Economy of European Social Policy*. Madison: University of Wisconsin Press.

Suleiman, Ezra N., ed. 1984. *Bureaucrats and Policy Making*. New York: Holmes and Meier.

Szyszczak, Erika. 1987. "The Equal Pay Directive and U.K. Law." In McCrudden, ed., pp. 52–73.

Trade Union Congress. 1947. *Annual Report*. London: Trade Union Congress.

————. 1950. *Annual Report*. London: Trade Union Congress.

————. 1965. *Annual Report*. London: Trade Union Congress.

————. 1967. *Annual Report*. London: Trade Union Congress.

————. 1970. *Special Report on Equal Pay to the Fortieth TUC Women's Conference: 1970*. London: TUC Publications.

————. 1973. *Conference on Equal Pay*. London: Trade Union Congress.

————. 1978. *Annual Report*. London: Trade Union Congress.

Vacca, Michele. 1992. *La Politica Comunitaria dell'Ambiente e la sua Attuazione negli Stati Membri*. Milan: Giuffré Editore.

Varillas, Benigno, and Humerto da Cruz. 1981. *Para una Historia del Movimiento Ecologista en España*. Madrid: Miraguano Ediciones.

Vogel, David. 1986. *National Styles of Regulation: Environmental Policy in Great Britain and the United States*. Ithaca, N.Y.: Cornell University Press.

Vogel, David. 1995. *Trading Up: Consumer and Environmental Regulation in a Global Economy*. Cambridge, Mass.: Harvard University Press.

Vogel-Polsky, Eliane. 1985. *National Institutional and Non-Institutional Machinery Established in the Council of Europe Member States to Promote Equality between Men and Women*. Strasbourg: Council of Europe.

Warner, Harriet. 1984. "EC Social Policy in Practice: Community Action on Behalf of Women and its Impact in the Member States." *Journal of Common Market Studies* 23: 141–67.

Weber, Max. 1992. *General Economic History.* New Brunswick, N.J.: Transaction Publishers.

Weir, Margaret, and Theda Skocpol. 1985. "State Sturctures and the Possibilites for 'Keynesian' Responses to the Great Depression in Sweden, Britain, and the United States." In Evans et al., eds., pp. 107–63.

Weir, Margaret. 1992. *Politics and Jobs: The Boundaries of Employment Policy in the United States.* Princeton: Princeton University Press.

Zabalza, Antoni, and Zafitis Tzannatos. 1985. *Women and Equal Pay.* Cambridge: Cambridge University Press.

Ziller, Jacques. 1988. "Conclusions and Issues of the IVth Erenstein Colloquium." In Siedentopf and Ziller, eds., pp. 130–42.

Zingariello, E. 1980. "L'Uomo e il Territorio." In de Paz et al., eds., pp. 237–42.

Index